THE LAST STRAW

THE LAST STRAW

A Continuing Quest for Life without Disposable Plastic

BRYANT HOLSENBECK

To Amy
w/ gratitude for
your attention &
creativity
Bryant

RCWMS | DURHAM, NORTH CAROLINA | 2018

THE LAST STRAW:
A Continuing Quest for Life without Disposable Plastic

Designed by Bonnie Campbell
Printed in the United States of America

ISBN: 978-0-9960826-6-2
Library of Congress Control Number: 2018943373

Copies of this book may be ordered from:
RCWMS
1202 Watts Street
Durham, NC 27701
www.rcwms.org
rcwmsnc@aol.com

All photographs by Bryant Holsenbeck unless otherwise noted.

For Ann Woodward and the Scrap Exchange,
who for the last twenty-seven years have kept close to three million pounds of plastics from going to the landfill and have inspired generations of artists and recyclers.

CONTENTS

PART TWO

THE LAST STRAW

INTRODUCTION

March 2018 was "No Straws Month" in Durham, North Carolina, by proclamation of our mayor, Steve Schewel, and with the encouragement of the citizens group, Don't Waste Durham. My car has two "Green-To-Go" reusable, take-out food containers in it, both ready to be returned to one of over twenty locations throughout the city so that I am ready the next time I am at the food co-op, or the delicious barbeque joint, Q Shack, or one of the growing number of local participating restaurants where I can get my take-out order or leftover food home without making more garbage.

In the almost ten years since I began using as little plastic as possible, I have noticed that more and more people are paying attention to our overuse of plastic. This spring the Church of England suggested that its congregants consider giving up single-use plastic for Lent. My Facebook feed continues to flow with stories of places where single-use plastic has been banned. California has banned plastic bags. France has banned plastic dinnerware—cups, plastic forks and spoons, the lot.

This battle for awareness, for care, for protection of our precious natural environment and its resources may not be won in my lifetime, but in the ten years I have been paying attention, the tide seems to be turning. Saying "NO" to single-use plastic is becoming much more commonplace. Just today I read about a Dutch grocery store that will be offering plastic-free aisles.

This book tells the story of how I learned to live differently, to become more aware of the world around me, and to understand more about plastic and its impact on us and our environment. It is based on the blog I kept when I decided at the beginning of 2010 to try to live without plastic for a year. If you are reading this book, then you are probably on your own environmental journey, or ready to begin. These days, living with as little single-use plastic as possible is an active habit for me. I am not as strict with myself as I was during that first year, but I keep a mental tab about where I can buy foods in bulk and which restaurants use throw-away cutlery and plastic plates. I don't return to the latter. For me, this has become a way of life. I try to do the best I can each day and not blame myself for being imperfect.

Welcome to the journey where together we are discovering what it means to live without single-use plastic. You are in good company as you say no to plastic straws, plastic shopping bags, and other useless plastic. You are in good company each time you re-use something rather than buy it new and when you recycle rather than throw away. You are in good company each time you pick up the litter others have left behind, whether it is at the beach, on a walk in the woods, or in your own backyard. May we all keep our eyes open and do the best we can as we care for the world around us and for each other. All the time.

—BRYANT HOLSENBECK
Durham, North Carolina, March 30, 2018

PART ONE

A YEAR WITHOUT PLASTIC

I. IN THE BEGINNING

January 1, 2010

Welcome to *The Last Straw: A Continuing Quest for Life without Disposable Plastic.* My name is Bryant. I am an environmental artist. I make art out of stuff no one wants any more. This year I am going to try an experiment: one day at a time, I am going to attempt to use as little single-use plastic as I can.

As an artist and a sculptor, I have been working with and using societal castoffs, the stuff we throw away, for many years. I have made works both using this stuff and that chronicle its quantity. I do a lot of work with students from elementary level to college age. In November 2009, I was invited to install one of my mandalas of thousands of bottle caps in the FedEx Global Education Center at the University of North Carolina. I worked with many students to make it and noticed that a good many of them were carrying their own refillable water bottles. It was a eureka moment for me when I realized that this was something that I wanted to do myself, and that by doing it, I would be taking a tiny step toward reducing the vast quantity of plastic that I had been thoughtlessly using in my life.

In my everyday work, I was working in many elementary schools throughout the state and had been struggling with how to teach students about plastic pollution. Whenever I asked young students where plastic comes from, they looked at me in consternation. Usually, they simply did not know, or had not thought about it. The two answers I got most often were:

1. from the store, and
2. from trees (like paper).

If I pressed further, I got an answer like, "from the ground." It was very unusual for students to know that it comes from petroleum, just like the gasoline we use in our cars. These interactions with children always left me frustrated with my inability to explain how the plastics industry affects all of us personally, globally, and environmentally. So, after a week of hanging out with college students who were motivated to work for more sustainable environmental policies, it occurred to me to give a life without single-use plastic a try.

I do not expect the impossible of myself as I attempt a year without throw-away plastic, but I feel ready to try. I cannot expect others to change unless I am able to. As Thich Nhat Hanh tells us in the poem below, we all breathe the same air and drink the same water. We are all part of the global currents that surround us. I need to believe that what we do as individuals does matter, and that on some level, what affects me affects you. So, I begin my year with curiosity and the hope that I can keep the poet within me alive as I move forward.

If you are a poet,
you will see clearly that there is a cloud
floating in this sheet of paper.
Without a cloud, there will be no rain;

without rain, the trees cannot grow;
and without trees, we cannot make paper.
The cloud is essential for the paper to exist.
If the cloud is not here, the sheet of paper
cannot be here either.[1]

. . .

One Sunday afternoon in November, with all errands run and books read, I was taking a walk on one of my favorite trails along the Eno River in my hometown of Durham.

The leaves were mostly down, so I was able to see the high sloping curve of the land to my left as I walked along with the river to my right. This is a much-traveled path for me. It is a path where yellow lady slipper orchids bloom each spring. Where and when to look for these elusive wildflowers is passed by word of mouth. The first time I saw them with my friend Fran, I remember thinking that from a distance their golden-brown color looked like sun bouncing off a shiny leaf, a patch of sunlight along the forest floor. Sometimes on this path closer to dusk, deer run deeper into the woods, white tails up. The path is clean, and it meanders beside a river that is also mostly clean. In the summer, I find plastic bottles and cigarette butts down by the swimming hole, but that's about it. These woods bring solace and joy to my life and serve as an antidote to the sorting and collecting of detritus which makes up a large part of my working life.

That November afternoon, as I crested the last hill on my way back to my truck, the thought came into my head: this coming new year, 2010, will be a year without disposable plastic in my life. I wondered, "Do I have the patience for this? What will I learn? Will I become a pedantic lecturer on the subject? How will my habits change?"

I decided to keep track by writing a blog about my experiences.

In December, a week or so before Christmas, my friend John helped me name my blog while we were having lunch at Rue Cler, a lovely French restaurant in Durham. I was drinking out of a glass with no plastic straw. I asked the waiter why I was not offered or given one (as is usually the case in most restaurants). "Well, I never assume anything," he replied. My friend laughed and said that my blog for the year should be called *The Last Straw*.

Since that time, I have been noticing everyday items that will give me pause once the year begins—the plastic wrapped around cheese and the top and straw on my cup from Wendy's as a friend and I hurried to the beach at Nags Head, North Carolina, to celebrate the new year.

This blog does not yet have a form. I hope that it will become a place to reflect upon the impact of my decision. I am optimistic that I will find ways to do without disposable plastics. The two bags that I have just collected, crammed full of bits and pieces of plastic that washed ashore here at Nags Head, are food for my mill. There is lots of plastic everywhere that we usually ignore. Does what I pick up matter? Am I picking up the last straws on the beach? It is easy to walk along and not notice. But for me, this year, my job is to notice. So here I go, on my last beach walk of this vacation, with a bag and a camera.

Happy New Year, everyone!

2. HOW LONG DOES LITTER LAST?

January 2, 2010

Here are the decomposition times for some common items:

Glass bottle: 1 million years
Monofilament fishing line: 600 years
Plastic beverage bottles: 450 years
Disposable diapers: 450 years
Aluminum cans: 80–200 years
Foamed plastic cups: 50 years
Tin cans: 50 years
Leather: 50 years
Nylon fabric: 30–40 years
Plastic bags: 10–20 years
Cigarette butts: 1–5 years

Wool socks: 1–5 years
Waxed milk cartons: 3 months
Apple cores: 2 months
Newspapers: 6 weeks
Orange or banana peels: 2–5 weeks
Paper towels: 2–4 weeks[2]

The list above is part of a longer one my friend Hunter Levinsohn compiled for a project that we worked on together for the Town of Chapel Hill a year or so ago.

Another piece of data: Taxpayers spend nearly $11 billion per year on cleaning up litter across the United States, ten times more than we spend on trash disposal. The estimated cost of litter pickup is thirty cents per piece of litter.[3]

This is day two in my yearlong experiment.

I am not pretending to be a scientist or a judge here. I do have a compost bin, and, like many Americans, a house with too much stuff in it. I spent most of this morning cleaning and sorting and making a big pile for the thrift shop. Phew!

Yesterday, I made it through two parties and a restaurant dinner without using plastic silverware or a plastic drinking straw. I remembered to tell the waitress not to give me a straw, and she remembered back. As for the parties, I got out my trusty travel silverware to use in place of the plastic silverware offered by the hosts. Didn't cowboys, at one time, carry around their own plates and bowls?

At party number one, Frank Phoenix encouraged me to read *Cradle to Cradle* by William McDonough and Michael Braungart.[4] The book jacket says:

> "Reduce, reuse, recycle," urge environmentalists; in other words, do more with less in order to minimize damage. But as architect William McDonough and chemist Michael Braungart point out in this provocative, visionary book, such an approach only perpetuates the one-way, "cradle to grave" manufacturing model,

> dating to the Industrial Revolution, that creates such fantastic amounts of waste and pollution in the first place. Why not challenge the belief that human industry must damage the natural world? In fact, why not take nature itself as our model for making things? A tree produces thousands of blossoms in order to create another tree, yet we consider its abundance not wasteful but safe, beautiful, and highly effective.

So now I have a book to read and a new paradigm to consider.

Today, I went to the Durham Farmers' Market and saw that it is easily possible to buy any vegetables I want without using plastic bags. Portia McKnight, farmer and cheesemaker, volunteered to wrap any cheeses I want in paper; all I have to do is call her in advance. This is an excellent discovery, as the cheese thing has been worrying me.

Today, I am feeling like I have given myself a yearlong homework assignment and am not quite sure what the point is. Unless perhaps it is to train my eye to identify what is biodegradable and what is not. Happy New Year, all! May your closets be clean and your compost bins cooking.

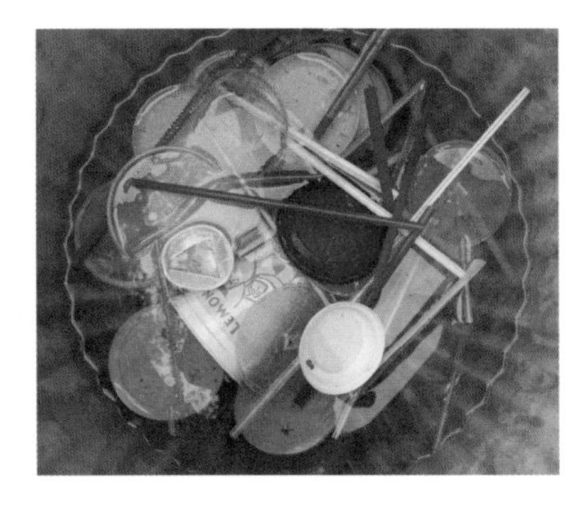

3. THE PLASTIC I LOVE

January 6, 2010

The beginnings and endings of all human undertakings are untidy . . .
—JOHN GALSWORTHY, *Over the River*

PLASTIC I LOVE:

1. My hot water bottle
2. My laptop computer
3. My cell phone and digital camera
4. All the synthetic parts of my running shoes
5. My contact lenses, reading glasses, and sunglasses
6. My baking spatula that gets all the batter out of the bowl

This is just the beginning of a very long list.

Needless to say, as I have been looking more closely at the plastic in my life, I see again and again how much I use on a daily basis.

On day six of this new year, I am still running on the fumes of the plastic left over from last year: a container full of yogurt in the refrigerator, cheeses wrapped in plastic.

How will things change when I finally run out?

On Sunday, I stopped by Lowe's on the way home from church to buy birdseed and found that every single bag of seed was plastic. What a disappointment. My friend Chris encouraged me to keep looking. So yesterday, in my local lawn and garden store, I found a fifty-pound sack of birdseed in a heavy paper bag. I felt vindicated; the search was worth the effort. Then I got home and ripped the bag open to find a plastic liner.

I am beginning to see that bits and pieces of plastic are absolutely everywhere.

Almost one week into this, and I am only slightly grumpy. I know I am still living off the leftovers of my life with plastic. Here are a few things that have changed already: I am not accepting plastic bags when I buy something. I am stopping waitpeople from giving me a plastic straw in restaurants, and I am avoiding plastic silverware. These have been fairly easy to date. It is the packaging around things I want to buy that is challenging me.

So far, the rules I know I can keep are:

1. Always carry my own re-useable bags and remember to tell clerks, "No bags, please." In many stores, I have to be vigilant and say this again and again for each purchase. Argh.
2. No plastic water bottles or Styrofoam or plastic cups. I now have two steel water bottles that travel with me.
3. No plastic silverware. I have decided to carry a knife, fork, and spoon with me. For my own enjoyment, I am using my mom's beautiful Lily of the Valley pattern sterling silver flatware.

I have much to discover and much to learn. Stay tuned for my thoughts on plastic water bottles and meditations on my observations, successes, and frustrations in grocery stores. Also, I want to know more about all of the plastic bags I am seeing in grocery stores that people say are biodegradable. How does this work? I want to know.

4. PLASTIC ON THE BEACH & CHOCOLATE CHIPS

January 12, 2010

I am twelve days into this year without disposable plastic. I continue to face challenges but am still determined. As I go about my world, small things I enjoy that are wrapped in plastic keep coming to mind. Stuff I am going to do without this year includes: most domestically made candy, because it is all wrapped in that silver plastic stuff, and breakfast or snack bars. Forget it. They are covered with the same kind of Mylar wrapping as candy bars.

This means I will be eating healthier snacks or maybe expensive, mostly imported, chocolate. This is certainly an adjustment I can make. (I will just try never to think about Reese's Peanut Butter Cups.)

Then there are the little triumphs. I finally found raisins in bulk at Whole Foods, and they were even on sale!

Today, when I was swimming, chocolate chips came to mind. I love to bake, but this will not be the year for chocolate chip cookies, unless I can find chocolate chips in bulk as well.

I keep thinking that all of this plastic around things must be about keeping stuff fresh as long as absolutely possible and extending shelf life. So why am I bothering to avoid it? I walk by the bags of plastic that you see pictured at the beginning of this chapter whenever I enter my home. They were gathered in two easy walks along the North Carolina coast (maybe an hour of walking). All of these bits and pieces of plastic were washing in and out with the tide—broken bits, and none of it food for any of us, ever.

My friend Rebecca Currie[5] suggests reading a book she owns about how people covered and contained stuff in "olden times." I am not ready to go there yet. My refrigerator is my friend, but Rebecca has a point.

This weekend, I am making my own granola out of bulk ingredients that I am collecting. It seems that most cardboard food boxes have plastic bags inside.

Shelley Wiley

5. HAITI HAS PROBLEMS

January 17, 2010

I've had an up-and-down week trying to dodge disposable plastic, which, in my bad moments, I feel is covering the earth.

All this week, I have been listening to the news of the earthquake in Haiti. For such a vulnerable country to be hit with a disaster of this magnitude is a hard thing to take in. What can we do to help, safe in our homes and towns where we have plenty of drinking water and easy access to hospitals?

I called my good friend Miriam Sauls, who has close connections in Haiti. For many years, she and her church have been working with a home for boys in Port-au-Prince. She reports that the main building is totally destroyed. Fortunately, everyone escaped alive. She says the organization has another building up the mountain where children with disabilities live. Part of that building is still standing, and all are living there as I write this. Miriam is feeling very fortunate for everyone's safety.

"What can we do?" I asked her.

Miriam seems to think that the main crisis right now is medical. Of course, Haiti will need long-term support for its recovery in the coming days, weeks, and years. If you have read the amazing book *Mountains Beyond Mountains* by Tracy Kidder, then you know about the important medical work of Partners in Health, an organization that works to bring modern medical care to poor communities around the world. The organization has three goals: "to care for our patients, to alleviate the root causes of disease in their communities, and to share lessons learned around the world."[6] Started by Paul Farmer, this organization has been working on the ground in Haiti for over twenty years and is one of the most respected nonprofits in the medical field.

Miriam sent me information about The St. Joseph Family, the organization that runs both the home for boys in Port-au-Prince whose home collapsed and the home for children living with disabilities up in the mountains, which was damaged.

The St. Joseph Family is supported in the United States by Hearts with Haiti, a nonprofit organization. (See www.heartswithhaiti.org.) The St. Joseph Family was started twenty-five years ago to bring five children off the streets and offer them a Christian family life. Now, there are three homes serving more than eighty children. Graduates from the first home, St. Joseph's Home for Boys, run a home for disabled children called Wings of Hope outside Port-au-Prince, as well as a home for young boys called Trinity House in the small coastal town of Jacmel.

The leadership of all three homes, and the boys themselves, believe in giving back in gratitude for having been brought off the streets. The St. Joseph Family has reached out beyond the immediate family to provide opportunities for the poorest children in their neighborhoods. There are now thirty-two girls from a nearby slum in Port-au-Prince attending school, eight neighborhood children participating in the community art center, seventy-two children attending the Lekol Sen Trinite day school, twenty-five children in the program in Jacmel, and five day students at Wings of Hope.

The building that collapsed during the earthquake served as a guest-house and dance theater, which were the Family's two main sources of revenue. So now, not only are the twenty boys who lived at St. Joseph's homeless, the revenue, which supported the extended family, is no longer available. They are in desperate need of help, with literally hundreds of people directly or indirectly depending on them.

The Wings of Hope home for disabled children is also seriously damaged, except for one room where all the children and caretakers are huddled. Care for these children is very supply-intensive, and the supplies are rapidly running out.

6. MY COUSIN MONTY'S GRANOLA

January 18, 2010

During the past week, I managed to collect most of the ingredients for the granola recipe posted below. This is my cousin Monty's recipe, and when she gave it to me four or so years ago at Christmas, pasted on a beautiful glass jar full of granola, I was in heaven. Finally, I tried to make it myself. Mine was great, but Monty's is even better. I think the difference was that I used honey, and she used maple syrup. I am only guessing about the difference; mine, according to my friend Cici, is terrific. I have been eating and eating it.

Why granola, anyway? I discovered that the bags inside nearly all cereal boxes are plastic. They feel a bit like waxed paper, so I had never really thought about it before. The bags are *not* biodegradable. Realizing this, I thought of Monty's recipe.

I bought all the ingredients from the bulk bins at Whole Foods, except wheat germ, canola oil, and honey. Those I was able to purchase in glass containers.

One word of caution! Do not be overgenerous with the flax seed, because you will spend the rest of the day picking it out of your teeth. I had more than the recipe called for and added freely. Though I am sure I will be very healthy after eating this batch, in the next one, I will use less.

Before I sign off this entry, I want to mention my friend Rebecca Currie's blog, again.[7]

Recently she told me: "My project this time around is easy—healthy food (whole grain, fruits, vegetables) for $100/month, shopping mostly at Whole Foods. Slightly more than I usually spend, though I'm not always so healthy."

I will be following her progress. I am seeing that eating without disposable plastic means eating fresh (before the stuff gets into a box) as often as possible. I had not really thought about this when I signed on to this project. Like Rebecca, I am not always so healthy, so this might be an unexpected bonus of my project.

Here's the recipe.

MONTY'S GRANOLA

2 cups old-fashioned rolled oats
1 cup raw sunflower seeds
1 cup raw sesame seeds
1 cup raw pumpkin seeds
1 cup raw flax seeds
1 cup raw wheat germ
1 cup oat bran
1 cup dry milk powder
1 cup maple syrup or honey
1 cup canola oil
1 tablespoon vanilla extract

Mix dry ingredients in a large bowl. Mix oil, maple syrup or honey, and vanilla in a microwaveable container and heat the liquids (60 seconds in a microwave).

Mix all ingredients very well. Place on two cookie sheets with rims. Bake at 225 degrees for about 2 hours. Stir every 15 minutes. Bake until golden. Thanks, Monty!

OPTIONAL: during the last 15 minutes of baking, add dried cranberries, raisins, banana chips, and/or any nuts of your choice.

7. PLASTIC ISLANDS: OUR OCEANS ARE US

January 23, 2010

> *A small group of thoughtful people could change the world. Indeed, it's the only thing that ever has.*
>
> —ATTRIBUTED TO MARGARET MEAD

Saturday—ah.

In a short while, I will be off to my local farmers' market to buy some very good bread and goat cheese, both of which will not be wrapped in plastic. This has been a hard week for me, plastically speaking. I have been working in an elementary school with some lovely fifth graders. They are smart kids, and it has been engaging to work and talk with them. They are interested in the world, and we have worked well together.

Once again, as I talked with them, I realized the ubiquity of plastic. We had the usual conversation. I asked, "Where does plastic come from?" The kids, usually so willing to answer questions, just looked at me and then gave answers like "trees" or "plastic comes from plastic." Eventually, we got around to figuring out that plastic comes from petroleum.

I was reminded yet again that plastic is an integral part of our everyday lives. If you spend a minute in almost any school cafeteria, you will see plastic silverware wrapped in plastic, Styrofoam soup cups, and disposable packets of mustard, ketchup, and mayonnaise.

Last year, I knew all this but was ignoring it, saying "It's not going to change anytime soon." My job now is to observe it. Last week, I ate my hot dogs without mustard and did not complain.

I showed the students the pictures of the islands of plastic our oceans are collecting. "The ships make the waste," the kids said in response, and that is partially true, but in reality, all that trash by the side of the road, all of those plastic bags floating down the road, all of those Styrofoam cups rolling around—they go into our streams, which go into our rivers, which go into our oceans. I have been thinking about this since Oprah talked about it on her 2009 Earth Day show.

The next time you take a walk anywhere, if you bother to look around a bit, very quickly you will see some shiny clear plastic that covered a straw, or a plastic toy from a fast food chain, or a piece of candy. Just look, and you will see it. It is there.

If you have gotten this far, thanks for listening to me.

8. RECONNAISSANCE

January 29, 2010

This week, I have been traveling to Raleigh early each morning to work at Raleigh Charter High School. On the way home, I try to stop at stores to look for food without plastic around it. Aldi, where stuff is cheap and you have to pay twenty-five cents to use a shopping cart, was a complete bust. I bought an avocado and some very bad ice cream. Trader Joe's was not much better. However, there were some apples, eggplants, and bananas without plastic coverings, and I was able to find a good deal on maple syrup in a glass bottle. I asked a young man about chocolate not wrapped in plastic, and he was very helpful, finding me some Swiss chocolate wrapped in paper and aluminum foil down on a bottom shelf.

What is becoming clear is that my year without disposable plastic will be possible mostly because I live in a town with a very good year-round farmers' market, a Whole Foods grocery, and the wonderful Weaver Street Market, a co-op grocery with three stores in my area within driving distance.[8]

I will be eating locally, before the plastic wrap gets on the food. Thank you, Barbara Kingsolver, for your excellent model! If you have not read her book, *Animal, Vegetable, Miracle* (which chronicles her family's quest to eat local food for one year), then drop what you are doing and read it now.[9]

When people ask me how I am doing in my attempt to get rid of disposable plastic in my life, I tell them I am grumpy but determined.

Last spring, when I was staying in Elizabeth City, North Carolina, driving back and forth each day to a residency at Currituck County High School on the Outer Banks, I enjoyed views over estuaries and marshes and sometimes glimpses of our beautiful Atlantic Ocean. One afternoon, I was standing in the middle of a huge big-box store that seemed to have been recently planted in the center of a marshy green field. I was in search of sewing needles. As I stood in the middle of the store, looking for the correct aisle, I saw plastic lawn chairs, Styrofoam coolers, and much more. I thought to myself, "Most everything in this store, in fact almost all of it, is not biodegradable. The stuff we buy here will not decompose. The chairs will crack and break, the dishes will fade and get lost, the clothes will end up in bales of polyester sent to a third-world country. The packaging around most of the toys, tools, and food will get tossed in an instant. If we are lucky, all we can hope is that all this plastic will end up buried in a landfill and not floating in an ocean somewhere."

Later that week in Elizabeth City, I stopped at a small park to walk out onto a wooden pier. I had seen people fishing there every time I passed, and it looked like a great place to view the sunset. I parked in the gravel access area and walked out to the end of the pier, past cypress roots entwined with Styrofoam cups, beer cans, and shredded plastic bags. The panoramic view was of sky and water; the close-up view showed our thoughtlessness.

This fall, I was back on the coast, this time in Manteo, for another residency. For a brief time, North Carolina had an ordinance forbidding stores on the barrier islands from giving plastic bags to customers. Yes, there was a Walmart in Kitty Hawk that gave you paper bags instead of plastic ones!

This is a small answer to a huge problem, a tiny step.

Perhaps you heard about Bisphenol A, a toxin in most of the plastic used to cover our food in the United States. Europe outlawed this stuff long ago. How have we allowed this to continue to be used here? My friend Barbara, a wonderful nutritionist who works for my local Whole Foods, sent me information about what that grocery chain is doing to monitor and change this.[10]

Here is some good news: My friend Rebecca Currie (I keep mentioning her blog, *Less Is Enough*)[11] is still working on helping me find a way to make my own dish detergent. Also, I have found a local source of toilet paper that is not wrapped in plastic: Brame School & Office Products! It is nearby, and I can buy toilet paper for sixty-five cents per roll, with no plastic anywhere! And this weekend, I am finally going to make my own yogurt! Soon, I hope to have a few good recipes for you.

9. PAYING ATTENTION

February 1, 2010

There is no enlightenment outside of daily life.

—ATTRIBUTED TO THICH NHAT HANH

The title of this entry sounds Buddhist to me, or slightly like something one of my elementary school teachers might have said to her class in frustration: "Pay attention."

Paying attention is what I have been trying to do when I get near any sort of commerce. The minute I forget, voilà, I get plastic I did not bargain for. There are always, always, surprises.

A few weeks ago, on my way back from a day on the road, I exited the highway very close to my home. Tired and hungry, I spotted my local barbecue joint (no, I am not a vegetarian) and thought my life would be complete if I could only have a good ole North Carolina barbecue sandwich with coleslaw and hot sauce. "No problem," I thought, as I waited in the drive-through.

I would be fine ordering the sandwich wrapped in wax paper, and I knew not to order any wonderful Southern sweet tea, as that would surely come in a tall Styrofoam cup with a plastic top and straw. But when I reached into the white paper take-out bag, imagine my horror when I pulled out one of those Styrofoam clamshell boxes, with another plastic container of hot sauce inside, plus a plastic-covered plastic fork. I felt like a fallen woman, a true sinner. "How could I have been so dumb?" I kept asking myself.

This is about remembering and paying attention to what is in front of me. Perhaps if I had gone inside and spoken with the young woman taking orders, she would have obliged me by leaving all of that plastic out of my order. I am pretty sure she would have thought me a bit demanding, or at the very least, peculiar. One day I will find out.

As I write this, I am safe and warm in my own home. The world outside is covered with at least six inches of snow. Being snowbound has been such a luxury. I made yogurt for the first time. It was delicious. I made two batches of crackers. One batch was amazing, and the other tasted like a hundred little sawdust hockey pucks.

When I took that innocent walk in the woods last fall, and had the small epiphany about trying to live for a year without disposable plastic, I really had no idea of the complexity of what I had chosen for myself.

Earlier last year, after I had begun bringing my own bags to the grocery store, a young cashier said to me when, once again, I had forgotten my bags, "Ma'am, whenever I forget my bags, I always go back to my car or home and get them. That way, I trained myself to remember." His voice is one I have carried with me. This eighteen-year-old man is paying attention, and so should I.

I quickly trained myself to bring my own bags to the grocery store and have become good at spotting Styrofoam and plastic silverware before I come in contact with them. What I only slightly understood last fall when I made this commitment was how totally saturated the world is with plastic packaging. At that time, I had not noticed that the bags inside most boxes of cereal, cookies, and sugary snacks are plastic. I had not noticed that almost all cleaning products come in plastic bottles. The list continues. So far, I am still enjoying the ins and outs of figuring out how to navigate all this. Some days, I feel like Sisyphus. Some days, I feel the joy of discovery. I am also making new friends along the way. All of this for trying to pay attention to the plastic around us!

One thing more about this "paying attention" business. For some reason, I have begun observing the beautiful forms of the trees around me. As I drive here and there and find myself at a stoplight, or when I open my front door, I have been looking up and noticing the branches of the trees against the sky, their beautiful graceful forms. When I do this, just for a moment, life stands still. Nothing more or less. Just that.

10. WHAT KIND OF CHEESE DID THOMAS EDISON EAT? AND WHAT WAS IT WRAPPED IN, ANYWAY?

February 8, 2010

I have not failed. I have just found 10,000 ways that won't work.

—THOMAS A. EDISON

This past week has been an up-and-down journey. It seems my job is to figure out how to lead a life without disposable plastic and have a normal life as well. In short, I want to do this without letting it rule my life. There are books to read, movies to see, walks to take, and friends to visit and laugh with, and, if I am honest, way too many dishes to wash.

I have not been on the road or in a school this week, so I have had time to shop and cook. Even so, some days are frustrating. Below is a list of my week's successes:

1. Making yogurt was easy with a hot pad and a meat thermometer!
2. All toothpaste made by Tom's of Maine comes in aluminum tubes. All but the top is recyclable!
3. You can make your own brown sugar (yes, the bought stuff is all packaged in plastic) by adding 1 tablespoon of molasses to 1 cup of sugar.
4. In the same vein, powdered sugar can be made by adding cornstarch to white sugar and putting it in a blender. I have not tried this yet, as I don't know how cornstarch is packaged.
5. My Arm & Hammer unscented clothing detergent works very well for washing dishes. Who would have thought?
6. I made excellent chocolate chip cookies by chopping up a bar of Swiss Chocolate from Trader Joe's.

My special thanks to Portia McKnight of Chapel Hill Creamery, who cheerfully wrapped some of their excellent Hickory Grove cheese for me in wax paper. All it took was a phone call, and it was waiting for me at my local Saturday farmers' market.

As I said in my last post, I am only just beginning to realize how inundated with plastic we are. Interestingly, many problems, which have seemed overwhelming at first, have been solved relatively easily. And once solved, the answers have often been so clear, so upfront, and, it often seems, already known by many others, that I keep saying to myself, "Now, why didn't I already know or do that?" Tom's Toothpaste in aluminum tubes and the idea of making my own brown sugar are two good examples.

I keep getting surprising clues and affirmations. While waiting for *Crazy Heart* to begin at the Carolina Theatre on Saturday night, I began talking with a friend whom I had not seen in many years. Upon hearing what I was doing, she told me that she and her husband had been working hard for several years not to store any of their food in plastic, after hearing a program on *The People's Pharmacy* about Bisphenol A.[12] She thought she remembered that the show's hosts, Joe and Terry Graedon, do not buy any food packaged in plastic. Who knew?

In the spirit of Thomas Edison, I am continuing my investigation into this world of plastic. I have listed my recent successes over things which at one time seemed like huge obstacles (or at least irritating bumps in the road). I am leaving to do a residency in Jones County in a few days. The owner of the B&B where I will be staying assures me she will not give me any water in plastic bottles. All this to say, I am happy to find the world an accommodating place, as long as I remember to ask. Right now, I am off to try some Hickory Grove Cheese from Chapel Hill Creamery. Portia says it makes a terrific grilled cheese. I am so ready.

II. BOTTLE CAPS

February 16, 2010

It's logical to say that what I do is an act of faith. . . . It came to me. And I worked it out.

—WALKER EVANS

In 1999, I made my very first mandala out of bottle caps and jar lids. It was part of a show entitled *Hello Again*, curated by Susan Subtle for the new Tryon Center for the Arts. It was in the Bank of America atrium in Charlotte, North Carolina. Peter Richards, the artistic director of Tryon Center at the time, came to my home along with Susan for a studio visit when they were looking for North Carolina work for the show. I made a small mandala from caps and lids I had already collected out on my back deck to sell them on my idea. Peter looked at it and said, "It needs to be big, really big. Can you make one twenty-feet across?" "Yes, yes, *yes*," I replied. And my work began. My undergraduate degree is in sociology, not art. For several years, I had been collecting all of the caps and lids I used, along with those of helpful friends and neighbors. I kept thinking, each one of these caps or lids is a mark of food eaten, or something consumed, and the plastic ones are not biodegradable. By the time Peter, his wife Sue, and Susan Subtle came to my house, I had already collected a large quantity of caps and was ready to make my first mandala. They gave me the opportunity. I have not looked back.

Over the past ten years or so, I've made close to twenty installations out of my collection of caps and lids. At colleges and universities, museums, schools, in a mall, and once even in a synagogue, I have made them. I have installed them as far away as California and Maine. I always make them with the help of community members. My friends and neighbors have helped with the collection, and I must have over 100,000 caps and lids by now, most of which I store in a kind neighbor's shed. I use the same lids over and over again. In a way, you could now call me a bottle cap expert! I know that white is the most popular color for lids. I know that the most common lid is one of white plastic with a blue interior, off of soda and water bottles. I have a hunch, though, that the most common lid might be changing to a clearish color, like the ones on most water bottles now.

Sorting and using the lids over time, I have seen changes in what we consume, how we eat. I have also seen brand designs change. I now see many more orange sports drink tops than I ever saw ten years ago. Many caps that were once metal have changed to plastic. Mayonnaise is an example. The South's favorite mayonnaise, Duke's, has a lovely yellow and black cap. It used to be in a glass jar, but now only comes in plastic—both the jar and the lid. When will all my lovely metal Duke's lids become collectables? One thing I can say for sure: in the past ten years, I have seen fewer and fewer metal lids.

A week or so ago in yoga class, I told my friend Marcy how happy I was to have discovered a toothpaste which came in an aluminum, recyclable tube. She responded, "Oh, but the top is plastic." Yes, it is, and I expected nothing less. Even a lot of glass jars with metal screw lids have a clear plastic seal around the lid to make them tamper-proof.

Plastic is everywhere. I know this. It just is. This year, I am learning this over and over again. Sometimes my choice, if I am going to live any sort of enjoyable life at all (which, yes, I surely am), is to do the best I can. I choose the way with the least amount of plastic possible and then proceed.

Being on the road is the hardest. I have learned that my steel water bottles are a necessity. Thank you to Subway for vegetables and sandwiches wrapped in waxed paper. Most school cafeterias are impossible—Styrofoam, Styrofoam, Styrofoam.

I have been living *The Last Straw* life for six weeks. I have more questions than answers, more curiosity than frustration. My life is not plastic-less. It cannot be. However, it is a life with much *less* plastic in it, and for that I am happily grateful. As my friend John Morrison says, "It is better to do the best you can than to be paralyzed by the pursuit of perfection."

Now, isn't THAT true?

Special thanks in this post go to poet Kate Greenstreet, who I was lucky to hear read at David Need's house the other night. She gave me the Walker Evans quote that begins this post. If you get a chance, check out her website. [13] You will not be disappointed.

12. PLASTIC: THE FABRIC OF OUR LIVES

February 26, 2010

PLASTIC
noun
1: a plastic substance; *specifically* : any of numerous organic synthetic or processed materials that are mostly thermoplastic or thermosetting polymers of high molecular weight and that can be made into objects, films, or filaments; 2: credit cards used for payment—called also *plastic money*[14]

My father was a textile engineer. In one of my earliest memories he took out a magnifying glass and showed me what his suit jacket looked like up close. Under the handheld glass, the scratchy gray fabric was a jungle of color and twisted fibers. Another world, right there on my father's arm. Years later, I remember him bringing home a coil of nylon rope from work, or "Caprolon" as Allied Chemical was calling their version of nylon at that time. "Look," he said, showing me a tightly wound and shiny white loop of rope, no bigger in diameter than my young index finger. "Look," he said, "This rope could pick up a car!" When I thought of the clunky black '58 Chevrolet Impala in our driveway, and looked at that dinky rope, I thought, "Ummm, sure, Dad. Are we talking miracles here, or what?"

My father was on the road most weekdays, traveling to New York, South Carolina, and Virginia to visit Allied Chemical offices and factories. He was always arriving home with something new and peculiar. One spring Friday, he came home with two pairs of nylon golf pants, one a shiny neon green and the other an incredible reddish orange. This was the late '50s and early '60s, when the bravest thing I had ever seen a man wear was a pair of madras shorts. One Christmas, we covered our tree with "nylon" angel hair—so different from the spun glass stuff. Words like "safe," "non-toxic," and "wonder product" were used. I think it was my junior year in high school when my dad's New York secretaries made him a stretchy red nylon Santa suit, with white nylon trim and a cap to match. That Christmas, hip-hop dudes had nothing on my father. The outfit was saggy, shiny, and tight all at the same time, draped over a portly, balding, middle-aged guy who had been called "The Head" in college because of the size of said feature. He toured the neighborhood and then the larger expanse of town crying "Ho, ho, ho! And nuts and fruits and candy for you!" to friends and neighbors, strangers, and the world.

While this is a story about my father, it is also a story about the development of polymers in the world, especially after World War II. This synthetic combination of materials brought us such things as nylon hosiery and carpets, polyester suits, and plastics galore for use everywhere. Polymers served to make everything cheaper, cleaner, neater, purer, lighter, easier. They were where it was at, and for most of us, it seems to still be where it's at. I Googled "Caprolon," and the page that comes up is about chemical properties. It has headings like "Caprolons: modified with fullerenes and fulleroid materials." To a non-scientist, this is ancient Greek, and my eyes glaze over looking at the page.

This I understand: I sit at a desk made of extruded wood covered with white Formica. I type on my plastic-encased laptop, plugged into my plastic-encased hard drive and printer. On my desk are stacks of (plastic) CDs and DVDs. This morning, I am wearing blue tights made stretchy by some wonderful polymer. The same goes for the cover of my down vest and the warm boots I am wearing, inside and out. I sit in comfort at my Naugahyde desk, wearing my vinyl fake-fur boots, able to communicate with the wide world from my mostly plastic computer. Truly, this is a miracle.

This year, as I attempt to dodge disposable plastic in every way possible, I have been pleasantly surprised again and again. Last night, I spoke with longtime friend Beverly McIver. When I told her what I was doing, she said, "You know, I never, ever use Styrofoam—I just don't." I am seeing on some fundamental level that people want to pay attention. We all benefit from the progresses of the world, what plastic has done for us, yet mostly once we notice its excess, we want to do something about it. I never know who will give me a good tip for where to find food not wrapped in plastic. Thank you, Rebecca Currie, for spotting those coconuts at Whole Foods! In fact, just as my dad brought home the miracles of plastics in the '50s, perhaps our new generations of moms and dads are bringing home solutions to our unnecessary use and abuse of them. We are beginning to pay attention.

13. LORI KERR

March 7, 2010

MINDFUL

Every day
I see or hear
something
that more or less

kills me
with delight. . . .

—MARY OLIVER, *Why I Wake Early*

My friend Lori Kerr died on Wednesday. For the past two years, I was lucky to have a studio right next to Lori's in an old church in Durham. Our friendship was about our art, which if you think about it (although I don't think we did), is about life. It usually took me until mid-afternoon to get to my studio, but Lori was always there, working, thinking, working, making.

After I had been in the studio for a month or so, Lori arrived one day with an oxygen canister attached to her nose. “The doctor said I am not getting quite enough oxygen,” she told me. Lori did not miss a beat. For the rest of the time I knew her, she used oxygen. When she came into my studio, she attached herself to a long plastic cord that went to the oxygen tank in her studio. Our lives there were about making art, about which we encouraged each other daily. With Lori, I did not think about that long plastic cord. I thought about animals and materials and ideas and just . . . life. Lori was one of the most creative people I have ever had the pleasure of knowing. She was a wonderful and loving mother to her children, and a good wife as well. She was an excellent friend. One day, she asked me to go on an urban chicken coop tour. I did not get to go, but I know her friends who did had a fabulous time. Lori was so busy living, loving, and creating that I didn't think she had time to die. On Wednesday, March 3, 2010, Lori Kerr decided to take herself off the ventilator that was keeping her alive. I am deeply saddened by her death. More than that, I am grateful for my time with her, her eye for animals, her creative mind, her living of each moment with curiosity and joy, her love for all of us humans and animals in this world. Thank you, Lori, for living so well. Your friends, your family, the lizards, the hedgehogs, the crows, the rabbits, the chickens, all of us—we love you and we miss you. You are a shining star.

14. PLASTIC AT THE BEACH

March 31, 2010

I know of a cure for everything: salt water. . . . sweat, or tears, or the salt sea.
—ISAK DINESEN, *Seven Gothic Tales*

Man had always assumed that he was more intelligent than dolphins because he had achieved so much—the wheel, New York, wars, and so on—while all the dolphins had ever done was muck about in the water having a good time. But conversely the dolphins believed themselves to be more intelligent than man—for precisely the same reasons.
—DOUGLAS ADAMS, *The Hitchhiker's Guide to the Galaxy*

I have been staying at the Atlantis Lodge on the North Carolina coast for the past few days. I hear the ocean wherever I am. It is a constant. My feet are cold from a morning walk along the ocean's edge. Earlier this morning, I was in the office answering my email, generally checking up on business, when I overheard one of the employees talking. "Yes, Fred and Ethel are back. I just saw them in the pool." I learned after asking her that Fred and Ethel were a pair of mallard ducks who had mated here last year and nested in the shrubbery near the motel sign, next to the road.

It is early spring along the North Carolina coast. I have seen lots of geese flying overhead and pelicans gliding along the surface of the sea. Except for Monday evening, the water has been quite calm. For the past few nights, around ten o'clock or so, I have watched the full moon wax and wane, reflecting a river of silver on the ocean, as far as the curve of the earth and beyond.

For the previous three weeks, I had been doing residencies with school children in Wake and Johnston counties in North Carolina. I talked with at least 500 children, and with about 200 or so, I made wild animals out of string, scrap fabric, wire, and ubiquitous plastic bags. It is rare for children to know where plastic comes from. I brought my lunch everywhere, as I know all cafeterias are full of Styrofoam. I keep thinking about the reaction of a young woman in one of the schools. When I was talking about plastic, asking about what they knew, where it came from, what we could or could not do with it, she just shook her head and said, almost to herself, "I can't worry about that, it is just too much."

The truth is, sometimes, I think that too.

Last weekend, I spoke with Joanne Andrews, an art teacher at Rogers-Herr Middle School in Durham. She told me how she and her students had been working with recycled stuff and how excited and interested they had been. She told me about when they saw some famous albatross photographs. "It is complicated," she said. Her students were quite troubled with the images of dead birds with their stomachs full of plastic. "Why is this stuff out there, and why is it not biodegradable?" were two questions on their minds. These are certainly the questions on my mind as well, as I travel here and there in my daily life, trying to dodge plastic.

When I walk the beach here, kicking shells, looking at the surf, I always see plastic stuff. This morning, first thing, I picked up a tattered plastic bag shimmering in the wet sand. I thought about how it would blend in with the ocean water and be easily swallowed by sea life. Three minutes later, on my way back, I found another bag almost in the same place, along with a few empty plastic creamer cartons.

While doing a residency on the Outer Banks last fall, I met a local woman who told me that every morning as she walked the beach she collected plastic. Over and over, again and again.

We are not going to stop using plastic. It is too much a part of our lives. I feel it every time I walk by a vending machine full of captivating junk food. I want to buy the candy inside. This time last year, I would have bought it and been delighted with my purchase.

We are not going to stop using plastic. I know this and am grateful for the plastic oxygen tubing my friend Lori used in the last year of her life.

We are not going to stop using plastic. It is embedded in our lives. It is an "inconvenient truth," and we enjoy the benefits and conveniences of it every day. Can we be more careful, more thoughtful, about this resource we have been given? As I sit here in the friendly office of the Atlantis Lodge on a sunny afternoon, at the beginning of spring, on the last day of a quick vacation, this is my heart's desire. This is my wildest hope and my life's work.

And now I am going to take a nap or read a book. I can't decide which.

I know this as well and I hate to admit it: I really, really miss plastic drinking straws. All the time.

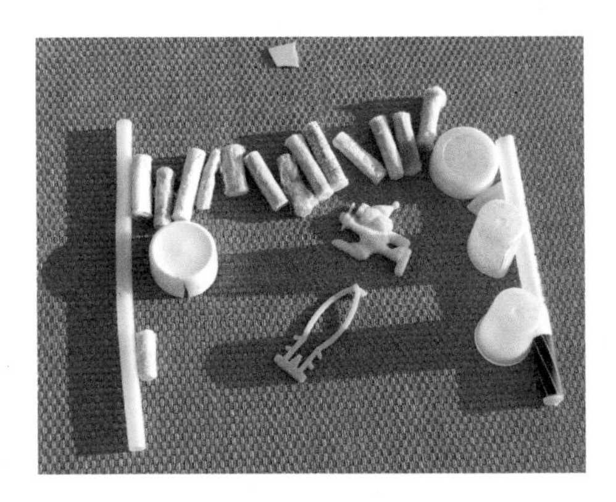

15. MULTIPLE CHOICES: "THE TIMES THEY ARE A-CHANGIN'"

April 11, 2010

Remember "The Times They Are a-Changin'" by Bob Dylan? Oh, it's such a good song. Driving to my studio yesterday, pondering the jumbled messages in my brain, I listened to a recent recording of Bob Dylan singing it at the White House. I thought I might title this post "Between the Idea and the Reality," with part of T. S. Eliot's "The Hollow Men"[15] at the top, but Bob Dylan won hands down. The recording is a beautiful version of Dylan's classic song. I heard in his voice sadness and hope. For those of us who woke up in the 1960s and '70s, we are all still here, and we are all still changing. For you younger folks, if what you're working for is making a difference, then let's go. We do not have time to waste.

In case you didn't read *Time* magazine this week, it has an article entitled "The Perils of Plastic," by Bryan Walsh.[16] The only big surprise for me in the article is that *Time* is writing about this at all. Walsh provides a very long and very complicated story. He believes we should be regulating the chemicals in plastics in the same way we learned to regulate pesticides and other toxins. This is such a good idea, so important and complicated to pursue. A quote from the article:

> As scientists get better at detecting the chemicals in our bodies, they're discovering that even tiny quantities of toxins can have a potentially serious impact on our health—and our children's future. Chemicals like Bisphenol A (BPA) and phthalates—key ingredients in modern plastics—may disrupt the delicate endocrine system, leading to developmental problems. A host of modern ills that have been rising unchecked for a generation—obesity, diabetes, autism, attention-deficit/hyperactivity disorder—could have chemical connections. "We don't give environmental exposure the attention it deserves," says Dr. Philip Landrigan, director of the Children's Environmental Health Center at New York City's Mount Sinai Medical Center. "But there's an emerging understanding that kids are uniquely susceptible to environmental hazards."

Next, have you, reader, seen Annie Leonard's documentary, *The Story of Bottled Water*?[17] If you have eight minutes and are the smallest bit intrigued, check it out. Did you know that we pay up to 2,000 times the cost of tap water when we buy bottled water? And often, when it is tested, we find that the water in these bottles is not as good quality as the water coming out of our very own American taps. Why is it that we are drinking all of this bottled water?

1. Convenience?
2. Habit?
3. Because we have been sold the idea that bottled water is good for us?
4. All of the above?

I see two things happening here. First, we are beginning to publicly acknowledge that plastic residue, coming to us through food packaging, may be harmful to us, and that we need to do something about regulating these chemicals. Second, we understand that maybe our habit of using plastics all the time might need to change—for us and for the earth. We do have choices. Paying attention to what we are eating and drinking, and how it is packaged, is a big one.

Okay—enough said here. Back to the personal. This week, I made a huge batch of Cousin Monty's granola and, of course, yogurt. Yogurt is so easy and so good from scratch. This year, instead of buying young plants, I am attempting to grow my own. Most annuals that I want to plant in my garden come in those plastic four-packs, but I am busy planting seeds that come in paper envelopes, curious about what will come up. Fortunately, I have lots of perennials in my garden. I hereby proclaim 2010 the year of "Pass Along Plants." I have plenty, if you need some and happen to live in Durham. I see dogwoods, azaleas, columbine, spring beauty, and, if I am lucky, a Jack-in-the-pulpit in my immediate future. I wish the same for all of you. Happy spring!

16. BEING CAREFUL

April 19, 2010

There is a river flowing now very fast. . . .
Know the river has its destination.
The elders say we must let go of the shore,
push off into the middle of the river,
keep our eyes open, and our heads above the water.

—ATTRIBUTED TO THE ELDERS, ORAIBI, ARIZONA HOPI NATION

This week, I have felt like I have been riding a river flowing very fast. Everywhere I am, I am surrounded by information, images, facts, and figures that tell me that I am right to be paying attention to the overabundance of throwaway plastic. I am listening, looking, and learning as hard as I can. It feels like sometimes all I can do is keep my head above water as I ride this rush of information that is swirling around me, like the Hopi Elders said.

What was I doing this time last year, when I did not constantly see single-use plastic everywhere as a huge and thoughtless waste of our resources?

Here is how the river flowed this week:

On Saturday, April 10, I saw *Waste Land*[18] by filmmaker Lucy Walker at Full Frame Documentary Film Festival in Durham. In this poignant documentary, Walker chronicles Brazilian artist Vik Muniz as he works with the "pickers" of Rio de Janeiro's *Jardim Gramacho* landfill, the world's highest-volume waste management facility. Vik photographs some of the workers, and then with their help, using the materials of the landfill, he makes transformative art out of stuff that people throw away blindly. In this amazing film, you literally see people riding the waves of garbage as they flow into the landfill, picking out the bits and pieces of recyclables that have been tossed away.

On Wednesday, I got to see part of Ian Connacher's film, *Addicted to Plastic.*[19] Connacher travels all around the world, talking to manufacturers and recyclers of plastic. He travels to landfills across the globe and to the Pacific gyre, one of the five huge ocean sites where much of our plastic is ending up. It breaks down into smaller and smaller pieces of floating bits, all of which attract pollutants and is eaten by wildlife.

My friend Amy Kellum emailed me a link to an article entitled "First, the Great Pacific Garbage Patch[20]; Now the Great Atlantic Patch." If you would like a verbal description of how the plastic detritus of our lives (plastic bags, broken toys, old water bottles, plastic silverware and toothbrushes, buckets, chairs, and who knows what else) is ending up in our oceans, breaking down to smaller and smaller bits of floating plastic, making a huge "soup" in our oceans, then read this article. Eighty percent of the stuff out there comes from mainlanders. jackson browne says this plastic pollution is a threat to our health and safety graver than global warming.

Speaking of jackson browne, I had the great pleasure of seeing him receive the Leaf Award from the Nicholas School of the Environment at Duke University on Saturday. This award is "given to an artist whose work has lifted the human spirit by conveying our profound spiritual and material connection to the Earth, and thereby inspiring others to help forge a more sustainable future for all."[21] I went to the ceremony because my friend Meredith Emmett sent me a notice about this award, and I had heard from friends how browne urges people to use non-disposable water bottles at all of his concerts. I was simply intrigued. What I saw was an amazingly intelligent poet of a man who has been looking for many years at how we have been wasting our resources and graciously urging us, with song and speech, to pay attention. He told us in his acceptance speech that single-use plastics leach chemicals, collect in our seas, gather toxins around them, become ingested by wildlife, and are uneconomical as well. I am paraphrasing here, but I was stunned by his clarity.

browne also said, "You have to be careful, because glass can break. You have to be careful, because the planet can break."

Since January 1 of this year, I have been looking for alternative, plastic-less ways to live my life. I have been using a lot of glass jars to store food in and to carry things around in. "Oops," I often think after an accident, "I know why people like plastic. It doesn't break when you drop it." I have to be careful.

17. STUFF

May 1, 2010

Stuff, stuff, stuff, and stuff. A shed of stuff, a studio of stuff, an attic of stuff, a house full of stuff, a mouse nest of stuff, a life of stuff. Once, when my friend Fran brought a new boyfriend over for me to meet, I introduced myself by showing him my big green roll-out trash bin full of stuff I had thrown away while cleaning the house for his arrival. "I really did clean up for you," I said. "Here is proof." I don't think he thought it was very funny, and anyway, he was history a long time ago. Somehow, though, because I work with detritus all the time, it is very important for me to let people know that I throw stuff away just like everyone else.

I have been making my living as an artist "documenting the waste stream of our society" (as my website intones) for quite a while now. I have no problem finding materials to work with. Stuff, stuff, stuff. The question for me is usually, "which stuff?" Furniture, shoes, clothes, packaging—the list of what is used for a short time and then discarded is endless.

When I give talks about my work as an artist and show images of my installations, one of the first questions people always ask is, "Where do you KEEP all of that stuff?" If they are asking this question, they are probably paying attention, themselves, to all of the "stuff" around them. Wherever I make an installation out of everyday detritus, I find custodians always, always get the point. Why? They are dealing with this "stuff" all the time—cleaning it up, sweeping it up, "disappearing" it into dumpsters.

Once I was talking to a class of elementary students as they looked at a bottle cap mandala that the high school students had helped to make. "Ms. Holsenbeck," the custodian, who had been watching me, said in frustration, "Why don't you just tell them to imagine how large the hole you would have to dig to bury all of this would be?" Good question, don't you think?

Meanwhile, the answer to where I keep all of this stuff is something like "wherever." In the shed, in the attic, in a friend's garage—wherever. Many of my friends have "Bryant Bags." These are usually brown paper standup bags

Michael Zirkle

with my name written on them, containing collections of credit cards, chopsticks, and whatever else said friend has saved just for me. I like this community effort; it has made my bottle cap mandalas, which are created from thousands of lids, the beautiful combinations of color and texture that they are. Let me make this clear—I am NOT a dumpster diver. I am a collector of everyday items that we use once and throw away. I am interested in their history—where they come from, how they were made, what they were used for, why they were thrown away, and finally the impact their waste has on us and the environment.

Stuff, stuff, stuff, everywhere, all the time. I do not think that it is a coincidence that since I have begun my reluctant year without disposable plastic, I do not want all of this stuff anymore, except maybe the credit cards and bottle caps. The tide is turning. I want less stuff around me. So I spent the morning with the intrepid Meghan Florian, emptying out the shed in the backyard. Bye-bye shoes with pithy sayings (remnants from old labyrinth projects). Bye-bye computer parts and plastic petri dishes and wooden boxes and little bits of this and that. They used to make up my sculptures, or baskets, or hats, or even installations. Some stuff went back to the wonderful Scrap Exchange from whence it came, some stuff to a friend who helps kids make musical instruments out of junk, the shoes to Ken Rumble and his gang of poets. Ken came over, loaded his pickup truck to the brim, and then used the shoes for his own installation performance. Yeah! And the rest to the dump.

Our world is just so full of stuff. Anyone who has ever cleaned out the home of an aging parent knows this. Anyone who has waited in line for their stuff to be weighed at the local landfill knows this.

Earlier in the week, I met with the sustainability committee of Guilford College about an installation I will be doing there in August. Sustainability. What does that mean? For me, sustainability means using our resources wisely, not buying stuff all the time (stuff we use once, stuff we don't need, stuff that isn't biodegradable, plastic). Mac McBee and Jim Dees showed me the huge composter they are using for food waste at the college. It is called an Earth Tub, a large, insulated, covered container with an auger that stirs the compost. It looks a bit like one of those giant Disney swirling teacups I always wanted to ride on, but bigger. For the first time, I actually saw the supposedly biodegradable corn cups becoming compost.

It was very exciting to see sustainability in action. Composting is a full-circle action. Organic material becomes fertilizer, which helps more organic materials grow. This is a simple, age-old concept. Are most of us in too much of a hurry to do this? Is our time so precious that we cannot stop and think about the consequences of our actions?

Do you have a compost bin? Do you have a garden? If you don't have either of these, do you have the time to think about what you are buying and consuming all of the time, and what happens to it once you throw it away?

18. APPLES

May 12, 2010

I am open to the guidance of synchronicity and do not let expectations hinder my path.

—ATTRIBUTED TO THE DALAI LAMA

In *The Hitchhiker's Guide to the Galaxy*, author Douglas Adams suggests that the answer to life, the universe, and everything is "42." These days, many friends want to know how I'm doing with my (okay, take a deep breath) "Single-use Plastic-free Year." Here's my report.

In a post dated January 6, 2010, I promised myself the following:

1. Always carry my own re-useable bags and REMEMBER to tell clerks, "No bags, please." In many stores, I have to be vigilant and say this again and again for each purchase.

2. No plastic water bottles, Styrofoam, or plastic cups. I now have two steel water bottles that travel with me at all times. I am proud to say that 132 days later, I still have those same two water bottles! Not losing them is a major miracle for me!
3. No plastic silverware. And no plastic drinking straws. I have decided to carry a knife, fork, and spoon with me at all times. For my own enjoyment, I am using my Mom's beautiful Lily of the Valley pattern sterling silver flatware.

By now, I am sort of trained in most of this. I try to remember to scan restaurants when I enter to make sure that they are not using throw-away plastic utensils, bowls, plates, etc. I remember to ask the waiter NOT to put a straw in my glass. At some places, I have worked out deals; one wonderful local family-owned restaurant lets me bring my own bowl. The other day, when I picked up my friend Miriam for lunch, I even remembered to bring a bowl for her as well.

I keep the space behind the driver's seat of my truck stuffed with reusable grocery bags. In the bags, I have smaller bags for bread, bottles, vegetables, things from the bulk aisle or the farmers' market. When in grocery stores, I usually head straight to the produce section and ignore most of the rest.

Traveling is tricky. Last week, while enjoying a residency in Camden, North Carolina, I made a peanut butter and jelly sandwich for myself for lunch each day, plus fruit and my water bottle. I do a lot of work in schools, and they always have cafeterias with lots of Styrofoam. Rather than be frustrated, I bring my own lunch. This week, I am working in Duplin County, which is a very large North Carolina county on the way to the beach with lots of farmland and many pigs (most of which I have not seen). I just ate lunch in a sweet restaurant in downtown Kenansville. I was served my beloved sweet tea over crushed ice in a glass. Last year, I might have eaten at the small restaurant that I saw on my way into town. The plate glass window was steamy, and the hand-painted sign on the side of the building said *Barbecue, Skin & Ribs*. There were lots of cars and trucks in the parking lot.

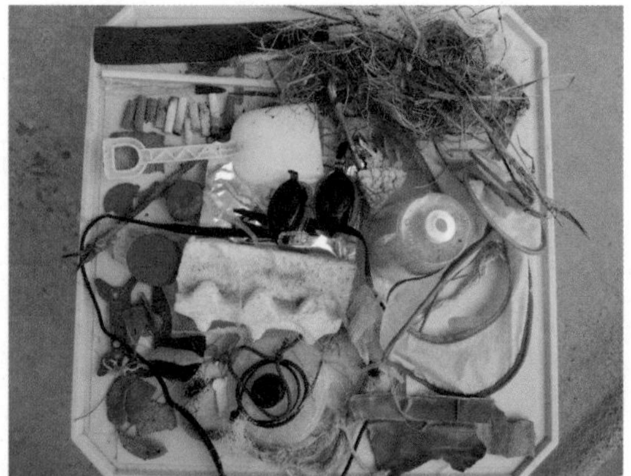

Just my kind of place, usually. However, betting on the Styrofoam present in most of these small establishments, I traveled on.

When I notice someone walking around with a large (non-recyclable) Styrofoam cup with a plastic straw poked through a plastic lid, I remind myself that this time last year, it would have been me. I guess you could say my view of the world has changed quite a bit since January 1.

I am surprised at how comfortable I am getting with using less plastic. Though I still miss plastic drinking straws, I am not missing packaged crackers, cookies, and cereal. My life is fine without processed frozen foods. I can buy both Edy's Ice Cream and Yum Yum Ice Cream (out of Greensboro, North Carolina) in cardboard cartons. If you are trying out Yum Yum for the first time, do not pass up the black cherry!

I am eating much less meat and cheese, and more eggs, bread, and vegetables. Because Weaver Street Market has chocolate chips in bulk, Toll House cookies are still in my life! If you want to try something really fabulous, try the yogurt-covered almonds in the bulk bin at Whole Foods.

There are still some dodgy issues. A lot of take-out containers are plastic. Cosmetics are mostly plastic-covered. I am still hunting for an affordable face cream in a glass jar. I have not run out of what I have yet, but neither have I given up the search. Many things I thought would be impossible to

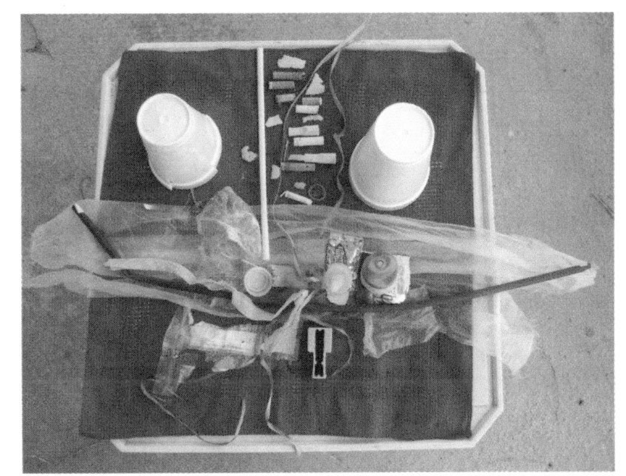

solve have been surprisingly easy. Making my own yogurt is an excellent example of something that turned out to be a simple and enjoyable process. It tastes great, and requires no plastic container!

So, the true answer when people ask, "How are you doing?" is, "Fine, just fine." I am constantly coming up with answers, solutions, and ways of doing without plastic. Mostly (except when I am very tired and possibly hungry), I have a very good time with this. When I get tired and grumpy, I just eat an apple, or maybe some nuts, and then I feel better. If Douglas Adams says the answer is the number 42, I think for me, right now, the meaning of life is apples. Pretty sure it's gotta be apples.

19. SPIN THE BOTTLE

May 23, 2010

Americans now drink more bottled water than milk or beer—in fact, the average American is now drinking around 30 gallons, or 115 liters, of bottled water each year, most of it from single-serving plastic containers. Bottled water has become so ubiquitous that it's hard to remember that it hasn't always been here. . . . Millions of Americans still drink tap water at home and in restaurants. But there is a war on for the hearts, minds, and pocketbooks of tap water drinkers, a huge market that water bottlers cannot afford to ignore. The war on the tap is an undeclared war, for the most part, but in recent years, more and more subtle (and not so subtle) campaigns that play up the supposed health risks of tap water, or the supposed health advantages of bottled water, have been launched by private water bottlers.

—PETER H. GLEICK, *Bottled & Sold: The Story Behind Our Obsession with Bottled Water*

When we're done, tap water will be relegated to showers and washing dishes.
—SUSAN WELLINGTON, president of the Quaker Oats Company's United States beverage division.

Driving home earlier this week, I heard Peter Gleick talking with Terry Gross on *Fresh Air*. Gleick is a MacArthur Fellow whose book, *Bottled & Sold: The Story Behind Our Obsession with Bottled Water,* has just been published. I learned that our tap water is very good; tap water and bottled water are regulated differently, and we do not require bottled water to adhere to the same standards as our drinking water. Another factoid—if the name of the bottled water has the name of a spring in it, then the water is probably actually patched together from a bunch of different springs. If the name is something else (like Dasani or Arctic), then it is tap water with the minerals taken out and the company's own special mix put back in.

Gleick says we cannot afford for all of us to only drink bottled water. It costs too much, and we are only recycling a small percentage of the PET (Polyethylene Terephthalate) bottles that it comes in. What is the deal here? What happened to our water fountains? Why have so many of us been carrying around those plastic bottles of water? I know, I know, I used to say this: "I keep filling it up with tap water." In and out of the car, the freezer, the refrigerator, again and again. I have certainly done this. I am just not doing it any more.

Here are two interesting news items.

Concord, Massachusetts, became the first US town to *ban*, yes, ban bottled drinking water. When I told this to a friend who works in the recycling industry, she forwarded an email entitled, "Duke provost ditches plastic water bottles." Apparently, after listening to jackson browne speak on the need to reduce single-use plastic when he received the LEAF Award at Duke last month, the provost took action. He said, "By switching from plastic bottles to a cooler and paper cups, annual drinking water costs are cut by $1,700—almost 80%. Not only that, but many employees have purchased water bottles and mugs to use instead of paper cups, cutting out even more waste."

Around twenty years ago, curbside recycling came to my town. In that first blue bin distributed by SunShares (a non-profit developed for neighborhood recycling), we were able to put our glass bottles and tin and aluminum cans. No plastic. There was not enough of a market for it. Gradually, some plastic bottles began to be collected—just the PET soda bottles and the HDPE (high-density polyethylene) milk jugs. These days, the plethora of plastics begging to be recycled is mind-boggling, and there is much conversation about it within the recycling industry. One thing to remember: recycling any of this stuff requires a market for it at the other end. Those numbers surrounded by the three arrows on each container are the plastic industries' label for what kind of plastic it is. More and more, our personal recycling should be about *paying attention* to what our local waste industries can handle, i.e., what they have found markets for. Even so, I feel it is important to understand that much plastic is down-cycled, as in it will be made into carpet or plastic lumber instead of the same thing again, like aluminum cans and glass bottles. This reminds me of William McDonough and Michael Braungart's book, *Cradle to Cradle,*[22] that challenges us to develop an economy where all our once-used goods are not thrown away, but reused in a valuable way.

As I write this (while looking at pictures of plastic bottles in our oceans, along our rivers, and in our landfills), I can't help thinking of the oil currently pouring into the waters of the Gulf of Mexico from a Deepwater Horizon/ BP oil rig. Are we the willing victims of our own industrialization? Oil is spewing from the depths of the ocean, coating our wetlands, and we cannot stop it. Plastic bottles are being tossed away, every day, all the time. Can we ask these industries to act responsibly, to make decisions that are sustainable for future generations? Will they hear us if we do? Gleick says we recycle less than 30% of our plastic bottles. Will rivers of petroleum in the ocean or petroleum's byproducts be more deadly to us in the long run? Will we be smart enough politically, scientifically, and personally to see these wake-up calls for what they are? Are we willing to take action and assume what responsibility we can? In my lifetime, I have been responsible for a *lot* of plastic bottles. How many of them did I actually recycle? I am not sure, but I know many were probably just tossed. I cannot be self-righteous here. No way. A friend has promised me that if I ever start pointing my finger at him, he is going to start calling me by the name of a crazy neighbor of his who has a conspiracy theory about plastic. So, I TRY to be careful.

Perhaps we all walk a careful line in this huge universe, which is spinning so fast, around and around and around. Forever. Or so we hope.

20. SAFE PASSAGE

June 9, 2010

What gives value to travel is fear. It is the fact that, at a certain moment, when we are so far from our own country, we are seized by a vague fear, and an instinctive desire to go back to the protection of old habits. This is the most obvious benefit of travel. At that moment we are feverish, but also porous, so that the slightest touch makes us quiver to the depths of our being. We come across a cascade of light, and there is eternity. This is why we should not say we travel for pleasure. There is no pleasure in traveling, and I look upon it more as an occasion for spiritual testing. If we understand by culture, the exercise of our most intimate sense—that of eternity—then we travel for culture. Pleasure takes us away from ourselves in the same way as distraction, in Pascal's use of the word, takes us away from God. Travel, which is like a greater and graver science, brings us back to ourselves.

—ALBERT CAMUS, *Notebooks 1935–1942*

Last night, I returned from two weeks in Guatemala. During my stay, I experienced the power of Tropical Storm Agatha, which roared through only two days after the eruption of the volcano Pacaya. The eruption left Guatemala City covered in black volcanic sand. Also, while in Guatemala, along with many industrious Guatemalans, I climbed over a storm-caused landslide on my way down from Lake Atitlan, about four hours from the city.

The purpose of my visit was to work with the children of Safe Passage,[23] an organization in Guatemala City. I was invited by Carlyn Wright-Eakes, who is in charge of Safe Passage's visual arts program. Through her, I met many dedicated volunteers, teachers, students, mothers, and administrators of this amazing place.

Today, my body adjusts to another time zone. I turn on the tap and know the water is safe for me to drink. I call friends to reconnect and settle in. My physical body readjusts to food I am used to and a bed I am familiar with. An industrious local teenager cuts my grass. Today, I am wondering about normalcy. I keep remembering a small Mayan woman, deftly climbing over the striated earth of one of the recent landslides in Guatemala that had blocked all traffic. She was dressed in the traditional *traje*—a beautifully colored woven skirt and embroidered *huipil*. Balanced delicately on her head was a huge bundle of calla lilies. Not missing a beat, up the wet brown-and-red earth, across, and down the other side she went. The sun was shining, laundry was drying, and the grass and crops on the mountainsides were a million shades of green.

We live our lives. We count our days. We are all traveling across waters great and small, along streets paved, cobbled, and pounded out of dirt. When we are lucky, we fly like birds above it all, and all this time, all we ever really want is to come home again.

Safety. A home. Isn't that all we really want? Isn't that why we fight our battles for education, for environmental justice, for food for all, for all of these passions of our hearts? All of the things we see and feel are important in this world.

Can you tell I am glad to be home? I give thanks for my journey and my own safe passage.

Stay tuned for tales of *mariposas* with the children of Safe Passage in Guatemala City, as well as musings on all of the single-use plastic I saw in my travels. Those are other stories for other days. Right now, special thanks to all of the wonderful people there at Safe Passage in Guatemala. Hi Carlyn, hi Liz, hi Carmencita, hi Gabe! Love to you, the rest of the staff, the kids, the volunteers, all of you, and thank you.

21. CAMINO SEGURO

June 17, 2010

You suppose you are the trouble
But you are the cure
You suppose that you are the lock on the door
But you are the key that opens it

—ATTRIBUTED TO RUMI

One day, Hanley Denning, the founder of Safe Passage, saw in her heart what needed to be done and did it. That very same week, Hanley visited the families who worked in the Guatemala City landfill. She sold her computer and car and, using some money she had in savings, opened the doors of Safe Passage (known in Spanish as *Camino Seguro*) by enrolling forty of Guatemala's poorest children in school. These children couldn't afford the books, school supplies, and enrollment fees required by the public school. This initial group received tutoring, healthy snacks, and care and attention that they so desperately needed. Another seventy children participated in a drop-in program, when they weren't working in the dump.[24]

Now, over ten years later, their programs include a *guarderia* (nursery school) for small children, a school for school-aged children, and an adult literacy program. After completing their degrees, many of the women in this program have been granted scholarships to pursue degrees in nursing or other professions.

Camino Seguro has a professional staff augmented by volunteers who come from all over the world to work there. "Today, Safe Passage provides approximately 550 children with education, social services, and the chance to move beyond the poverty their families have faced for generations."[25]

Every Thursday, they offer a tour of the landfill. It begins in a cemetery. The day I went, it was very rainy. We walked from the parked bus past graves large and small, until we stood on the edge of a cliff, looking down with the vultures into the landfill far below. Trucks rolled in and dumped their garbage while, from our vantage point, tiny ant-sized people, all covered in yellow plastic rain ponchos, swarmed the piles and sorted through it for items of value to sell.

In the United States, we pay workers to sort our recyclables in what is called a Material Recovery Facility (MRF). It is not a job I would want. Would you? Most of us just want our garbage to go AWAY. After all, we threw it AWAY, didn't we? When my friend Blair Pollock gives tours of North Carolina's Orange County Landfill, he begins by welcoming people to AWAY.

While I was in Guatemala, Carlyn Wright-Eakes, the art teacher at *Camino Seguro*, invited me to do an art project with the kids. Bringing "stuff" donated by the Scrap Exchange of Durham, and with the help of Bonnie Wright (mother of Carlyn) and Julia Gartrell (store manager of the Scrap Exchange), we made recycled *mariposas* (butterflies) with the children. On the tail of each *mariposa*, the children wrote their hopes for all of the Guatemalans suffering from the ravages of the recent landslides caused by Hurricane Agatha. The kids were joyous. The staff and volunteers were very helpful, as

were the teachers. The Safe Passage facilities are situated in Guatemala City near the landfill operations and the homes of the workers and their families. They are large and welcoming, with gardens, lots of light, and good food. It is a great place to be. Pictured above are some of the butterflies the children made. Their English was much better than my high-school Spanish.

I have been writing this year about my attempt not to use disposable plastic. I have discovered over and over again that it is everywhere. In Guatemala, people do not drink their tap water until they filter it. During my trip, I was mostly successful in not using plastics, but not always. Everybody in the United States likes to give you one of those filmy plastic bags for everything. Guatemala is certainly no different.

Experiencing Hurricane Agatha on Lake Atitlan and hearing booming thunder-like sounds and finding out that they were landslides—these were sobering experiences. After the storm was over, the sun came out and the people began digging the mud out of their homes, doing their laundry on the rocks, fishing along the edge of the lake, and salvaging firewood that had washed down into the lake. You could not help but see the plastic detritus that was floating everywhere.

While in Guatemala, I was able to see and experience just a bit of the amazing work done by *Camino Seguro.* Go to their website and you will be amazed as well. I was reminded in many ways of how vulnerable we are as human beings. What we give to each other every day really matters, no exceptions. Whether we are teaching school, or riding a bus, or cleaning up after a hurricane, our skin is only a thin membrane around our bodies, our bones, our brains, and our hearts. Yet we can do and affect so much if we put our hearts into it. Not only did Hanley Denning know this, she acted on it with all of her heart—*Camino Seguro* for all of us.

Special thanks to the Scrap Exchange and the fiber department of Penland School of Crafts for donating materials for our *mariposa* project and again, and always, to the people of *Camino Seguro*.

Also, thanks to my intrepid travel companions: Liz Love, Carlyn Wright-Eakes, Bonnie Wright, and Julia Gartrell.

22. CHECKING IN ON THE FLY

June 24, 2010

The knack [to flying] lies in learning how to throw yourself at the ground and miss.
—DOUGLAS ADAMS, *Life, the Universe and Everything*

When I came to the Virginia Center for the Creative Arts this week, I had plans to write post after post about plastic bags and other ubiquitous plastic byproducts of our lives. Not having to see, use, or deal with any of that stuff has been a big luxury. I am taking a break—just making art and trying not to think. Make, eat, swim, make, eat, laugh, walk, eat, sleep, eat, make, eat, swim, make, and then swim again. It has been a very good time.

Next week, when I'm back in Durham, the daily ins and outs of my plasticless year will continue. Meanwhile, you might want to check out the *Plas-*

tiki,[26] a boat made out of plastic bottles that is traveling around the world with environmental messages galore. These folks are paying attention to our overuse of plastic and doing something about it in the best possible way.

I will end this post with a poem by Metta Sama, a lovely poet whom I met here at the Virginia Center for the Creative Arts. She was the one who saw a red fox and fed the horses apples every night. She wrote this poem about the crows she saw me making out of recycled materiels. Thank you, Metta.

a murder of crows slicing the sky:
they can lead you to death or water,
depending on how you see them,
cloth stitched bodies,
albumed wings;
wings wide enough to cover a mountain of glass
or protect it from the burden of sun.[27]

—Metta Sama

23. WE ARE OUR COMPOST

July 9, 2010

The great obstacle to discovering the shape of the earth, the continents, and the ocean was not ignorance but the illusion of knowledge.

—DANIEL J. BOORSTIN, *The Discoverers*

I say beware of all enterprises that require new clothes.

—HENRY DAVID THOREAU, *Walden*

I am home. In the evening, a cacophony of katydids and cicadas surrounds me in the warm night air. The sidewalks are hot, and my garden droops at midday. The turtles are at the swimming hole, their heads just above the water's edge, peering at us humans with yellow eyes. At the farmers' market, I see blueberries, peaches, and all the tomatoes that a person could imagine.

I grew up in Greensboro, North Carolina, with a yard much like the one I have now. My father was often on the road, and when he was home, he gardened. There were vegetables in the back next to the fence, flowers in the side yard next to the driveway, and compost in a dip in the far-right corner of the backyard. On any given day, I saw grapefruit rinds, banana peels, and apple cores mixed in with the leaves and pine needles from our yard. Except for my college years, I think I have never lived without a compost pile; for me, it is part of home. My current compost pile is under a large magnolia tree in my side yard. It is a cylinder of half-inch-square metal fencing, about two or three feet across. I fill it with leaves, grass, cornshucks, peach peels, coffee grounds, and whatever else I have been eating. My technique is simple: Once or twice a day, I take my plant-based food waste to the compost pile and bury it. Once or twice a year, I remove the fencing cylinder from the pile, throw all of the new stuff that's on top back into it, shovel the rest into some five-gallon buckets, and spread it around my garden.

You might ask, "What does compost have to do with the story of your year without disposable plastic?" Everything, actually. As I have been listening, looking, watching, and thinking about this for over six months, I have noticed that whenever I speak with recycling professionals, at some point they *always* say the most important thing is to *compost*. They say something like, "Yes, it is good to recycle, but please don't throw those valuable bits of organic matter into the landfill where they will go to waste, instead of feeding the earth again and again." By composting, we complete the circle—birth, death, rebirth. Not only will this reduce the amount of garbage you throw away big-time, but it will make you feel good and put you back in touch with life's natural cycles. Worms are fun. Any kid can tell you that.

Over the weekend, I was working at the Orange County Solid Waste Management booth at our amazing trash-free Festival for the Eno here in Durham. A very well-meaning person earnestly wanted to recycle her yogurt tubs and could not understand why one municipality took them and another did not. (The answer is about markets and is not simple.) In the back of my mind, I kept hearing myself saying, "Compost, compost, compost." I ask you, if you are out there and you are an earnest recycler, then are you composting? If you are not, then why? Some of us come by it naturally, but if you did not, it is not that hard. My compost pile is just a bit of fencing. If you want more, compost bins are cheap. Whole Foods sells them for around $70. Orange County (North Carolina) Solid Waste has the same ones for around $50. My friends Karen and Randa bought a beautiful handy-dandy one with wheels, on a stand that spins and twirls, from local Durham-based artist George Danzer. When my niece was in school up in Cambridge, Massachusetts, she took her organic waste to a nearby Whole Foods. If you don't want to build or manage a compost pile, ask a neighbor if you can use theirs. For years, my friend Jenny stored all of her food scraps in a bag in her freezer, then stopped by my house every week or so to bury it in my compost. Solutions are everywhere.

There are also worm bins for apartment dwellers. You can do what Colin Beavan and his family did, as shown in the documentary *No Impact Man*[28]; it is not that hard. Afraid of worms? Find a small child, full of wonder, and they will help you out. The most popular I have ever been was during the year I worked for SunShares, a non-profit recycling organization (now defunct). I took a bin full of garbage-eating red wigglers into classrooms. Every kid wanted to touch them. You can too. You really can.

Here is what my friend Janine Faye from Massachusetts writes about her compost:

> My first compost was a pile of leaves and stuff from the garden, vegetable and fruit scraps and soil. My niece said "think parfait" layers. I poked holes in it and it all turned into usable compost in a year. It worked well for years. Even though I never added meat, dairy, rice, bread, or pasta, I was worried about attracting rats, so I got a closed black plastic container last year. It didn't break down as fast as the open pile, even though I added a lot of red wriggler worms from a neighbor's compost. I think I need to water it more. I used some of the stuff anyway and now there are eggshells in large pots holding tomatoes. A month ago, I moved the compost container and now use the not fully composted soil as one of the layers. It's really not that hard to do and very gratifying. It's out there doing its thing on its own. Just got to feed it and make sure there's air and water.

Thanks, Janine!

Plastics recycling is crucial and very complicated, with markets changing all the time. It is very important to pay attention to who takes what, because if you do not, then you are polluting a stream of plastic that is on its way to become something else. I will write more on this in a later post, I promise. For my friend with the #5 yogurt containers, I have two messages:

1. I know Whole Foods takes them, as I just saw a bin right in the front of the store.
2. At the end of this article, I am reprinting the yogurt recipe I published earlier in the year. It is very easy, very good, and has become a staple in my life this single-use plastic-free year.

Today, a recycling expert told me that 20–30% of what goes into our landfills is organic waste that could be composted. So, to all you composters and recyclers, I salute you and thank you for taking an active part in reducing our much-too-large waste stream. If you are not recycling, come on, now. If you are not composting, I know there is someone out there who wants to help you start. There are tons of websites, tons of books, and where I live, lots of good neighbors with room for more compost. If you think about it, compost is us, isn't it?

YOGURT MADE EASY (OR YOGURT WITH A HEATING PAD)

Recipe extrapolated from Harold McGee

Heat 1 pint (2 cups) of milk to 180–190 degrees. If you do not have a thermometer, this is when the milk starts to steam and tiny bubbles begin to form. Let milk cool to around 115–120 degrees. The milk will seem very warm to hot. Stir in one tablespoon of live yogurt, either from your last batch or borrowed from a neighbor.

Put this mixture in a clean warm glass container, wrap it in a towel, and place on a heating pad turned to low or in an oven with a pilot light. Ignore all day, or for at least 4 hours, and then voilà—you have yogurt. I have done this twice, once with whole milk and once with 2%. It was delicious both times. Also, I put this off and put this off, but it is VERY easy. I like a recipe that you can ignore most of the day.

24. AND LET THE COMPOSTING BEGIN

July 16, 2010

In California, they don't throw their garbage away—they make it into TV shows.
—WOODY ALLEN

Probably one of the most private things in the world is an egg until it is broken.
—M.F.K. FISHER, *How to Cook a Wolf*

So here is a question. How many compost bins do you think there are in Beverly Hills? Anyway, this just in from my friend Rebecca Currie, who writes the blog *Less Is Enough* and has been a huge resource for me in my attempt to live single-use plastic-free this year. For tips on eating plainer and simpler and just musings on life's curiosities in these crazy times, be sure to check out her blog.

After reading my post, "We Are Our Compost," Rebecca sent me the following:

> Sarah Cynthia Sylvia Stout, "A Guest Post by Rebecca Currie" (a.k.a *Less Is Enough*)
>
> I was talking on the phone the other night to a friend who started the year with a bunch of things she was hoping to accomplish, and she was feeling a little bit down, like she hadn't made much progress. She said she wrote out the list of everything she's done but when she looked at it, it didn't seem like much.
>
> I said, "Did you include taking out the garbage?" (She really hates taking out the garbage, and that was one of the things she wanted to do a better job with.)
>
> She laughed and said, "No! I forgot that. I'll have to add it." Then she said, "Did you ever read that poem in *Where the Sidewalk Ends* about the girl who wouldn't take out the garbage? Cynthia something Stout who wouldn't take the garbage out?"
>
> She said, "I'll add that to the list, that I didn't turn into Cynthia Stout and wasn't buried by my garbage."
>
> The next day I got an email from my friend with the text of the poem, and as I read it, I was struck by the fact that poor Sarah Cynthia Sylvia Stout wasn't buried by garbage at all, she was buried by food waste, and if she'd only had a worm bin this whole tragedy could have been avoided.
>
> And that made me think of my friend Bryant's last post, so I wanted to contribute this guest post for her, to reiterate what she said:
>
> **Reduce! Reuse! Recycle! Compost, compost, compost!** It will change your life. And spare you from the terrible fate of poor Sarah Cynthia Sylvia Stout, who would not take the garbage out.

Hello—It's Bryant again. I have to end this post by thanking Rebecca and telling you something amazing I learned about her the other day. After reading my post, she told me she actually only needs to take her garbage out once or twice a year. Wow, Rebecca—just wow.

This has got to be because of how carefully she shops and how much she recycles and uses her worm bin. And she is a great cook. She leads a full and interesting life with less garbage than anyone I know. Check out her blog, *Less Is Enough*, and you will be amazed as well.

25. YES, YES, YES: WHATEVER YOU CAN DO MATTERS

July 23, 2010

> *We are not going to be able to operate our Spaceship Earth successfully, nor for much longer, unless we see it as a whole spaceship and our fate as common. It has to be everybody or nobody.*
>
> —ATTRIBUTED TO BUCKMINSTER FULLER

It is hot and steamy here in North Carolina, as only July in the sultry South can be. My single-use plastic free-year is more than halfway over. I would like to think that I have been mostly successful in this experiment, if you give me a break here and there, like at the Durham Bulls ballpark last night. I scanned the food as I walked by all the concession stands on my way in and figured out I could buy a hot dog or a pizza without getting any throw-away plastic. And I had my steel H_2O bottle full of cold water. It was my godson

Theo's birthday, and a bunch of us, including his nine-month-old daughter, sweated and laughed and (sort of) watched the game. I was surrounded by plastic cups of beer, ice cream with plastic spoons, plastic bags full of cotton candy, and ice-cold slushies in Styrofoam cups. And yes, like the baby I was holding in my lap, I demanded my share of her mother Rachael's cup of orange Italian ice. I am not perfect, and it was so hot. Who could resist ice-cold bright orange crushed ice, even if it was in a Styrofoam cup? Certainly, neither of us.

When I think of all the single-use plastic that gets thrown away at one ballgame, I could ask myself, "Should I bother? What difference can one person make?" First and foremost, I am curious and amazed at absolutely how much of that stuff is out there. And second, I have to believe that paying attention to recycling and reuse makes a difference. If I did not, then I could turn myself into an apathetic blob of denial—boring, boring, boring. This year, I have noticed so much about plastic that I was ignoring in my rush through life. Also, and this is very clear, I am seeing and hearing all over the place that I am not the only one who is beginning to REALLY notice that we have a problem.

The other day, my friend Blair Pollock of Orange County Solid Waste told me that at least three countries (South Africa, Ireland, and Bangladesh), plus the city of San Francisco, have banned the use of plastic bags. Earlier, I wrote about the town of Concord, Massachusetts, doing the same. Why is this happening? Many complex reasons, for sure; the mess of non-biodegradable bags blowing and floating around everywhere is a big one. I have heard stories of rivers in India clogged with those cheap plastic bags, and I saw many similar signs of our overuse of plastic while on Lake Atitlán in Guatemala. Taking something out of the earth and making it into a material that has value for a very short time, then tossing it away, is simply not workable. Our world is big, and we have been able to hide from this for a long time. I am thinking our time for ignorance on this issue is about up.

The world of plastics has changed and keeps changing, all the time. I began this post wanting to speak about recycling plastics, telling what I have learned about which plastics can and which cannot be recycled. Googling the solid waste offices of the three municipalities closest to me, I got three entirely different and very specific lists. Why is this? Why aren't there any simple answers out there? I want to do the right thing, I really do. Why do *they* make it so hard? Really, I am busy here. This year, though, my job is to think about these questions.

Here is what I know:

1. The number surrounded by three arrows on all plastics is only an industry label as to what type of plastic it is. The three arrows *do not* mean it is recyclable. Is this confusing and misleading? *Absolutely.* Is the plastics industry interested in changing this confusing signage? NO, they are not.
2. The plastics we are able to recycle are the ones that our local municipalities (or in some cases, stores) have found markets for. Why is this different for each municipality? Recycling is about finding someone who wants to buy the product to use it again. These markets are not huge. They change, just as our desire for different plastic products changes.

3. In the past twenty years, plastics recycling has become extremely complicated. There are so many issues involved. We have made more and more things out of this cheap "throwaway" stuff. We are so smart that we change the chemical makeup all the time to suit our particular needs. It looks to me like the people of the plastics industry are not working together, but running a race to make as many different incompatible types of plastics as they can. When combined, they do not decompose and sabotage the chemical composition of each other.

All of this and more are what our recycling professionals are dealing with when they look for markets for our throwaway plastics. Blair Pollock of Orange County Solid Waste explained to me why his organization recycles what it does and does not. He said two very interesting things: (1) Those rigid clear clamshells that we buy fruits and vegetables in are made of many different types of plastic, many of which will pollute the big effort to collect PET plastic drink bottles. (2) All the plastics made of corn are only compostable in big industrial composters. They pollute the other plastics when melted with them, and you cannot compost them in a backyard compost pile.

Now I have to ask, "Why is everyone buying those corn cups and spoons?" To make ourselves feel better? Because you want to "do the right thing?" It could work. Guilford College in Greensboro, North Carolina, has an Earth Tub, which does compost them. Most people do not have these, nor do they have access to one.

The final complicated news: the plastics industry has developed oxo-biodegradable chemical additives, which will break plastic down into smaller and smaller particles. Is this actually helpful? It looks to me like the answer is No. What good is it to make plastic LOOK like it is gone, when it's not? In addition, these chemicals pollute the other recyclable plastics when they are mixed together.

I have written above about what I understand about the complexities of current plastics recycling. YES, I am an opinionated novice here, possibly on my way to becoming a ZEALOT. These are very complicated issues. All

I can say about doing the right thing (plastic recycling-wise) is to do your research. Find out what your town or county recycles, where they want you to recycle it, and in what form. Why should you do this? Does what you do matter? YES. We are all in this together. YES. Paying attention and doing the best you can matters. YES. YES, it does.

26. GRATITUDE

August 5, 2010

If the only prayer you said in your whole life was, "Thank you," that would suffice.

— ATTRIBUTED TO MEISTER ECKHART

As we express our gratitude, we must never forget that the highest appreciation is not to utter words, but to live by them.

—JOHN FITZGERALD KENNEDY

I figured out a while back that it doesn't take that much money to live well. I mean, really, why would I be an artist at all if I thought money would give me answers? I am fortunate in that I have been able to make my living working as an artist almost all of my adult life. "Where am I going here?" you may ask. "Isn't this blog about your struggle for a disposable plastic-free year?" Good question. The thing is, by now, halfway through the year, it is not really such a struggle. This is partly from educating myself about where and when to buy my food and how to dodge take-out containers. The other part is that I am used to living on a budget, figuring things out, being thrifty, shopping smart. I have done this all of my life. Making yogurt and granola, my own iced tea, or doing similar things to avoid plastic has not been a huge step for me. And I am RELIEVED to be dodging the big box stores, or as I now think of them, homes of plastic and plastic packaging. I do not need what they have. Over and over, I have thought, "Here is a place where I will have to use some plastic." Then, voilà, I figure out a way not to.

A week or so ago, I was planning a short camping trip to a lake up in Virginia. As I was gathering and borrowing equipment (a tent, cook stove, etc.), I thought, "Okay, I will need to buy plastic bags of ice, no choice." Fifteen minutes later, I was taking ice cubes out of my freezer and storing them in glass jars to put in my cooler. When I told my carpenter friend Dave this story, he said, "Where is the nearest store or restaurant with their own ice machine? Couldn't you just take your cooler in and ask them to fill it directly?" Duh! I kind of feel like "the little engine that could." If I just keep saying, "I think I can," or more exactly, "There is some way I can do this if I think hard enough," I usually can.

There have been a few recent surprises.

The butterfly bush in my garden, which grew and grew but never had flowers, turned out to be a peach tree that sprouted from a peach pit in my compost. This year, I harvested over twenty peaches. They were small, but very sweet. And the price was SO right.

Hurrying through CVS Pharmacy last week, wondering why I was there, I looked up and saw rolls and rolls of toilet paper wrapped in paper (not plastic film) on sale for fifty cents each. I just love it when big chains actually pay attention to the environment.

I am in love with Lush Cosmetics. This company comes from England and is doing a good job of being environmentally friendly. I have been buying terrific shampoo in bars from them this year. Plus, I can get moisturizer in bars and chunks of bath stuff that bubbles. None of these products is wrapped in shiny plastic film. No plastic on any of it. They do sell plastic pots of cream, which they will recycle if you take them back to the store. This plastic-free year, I am not going near those, but it is such fun and a relief to buy their other products.

And finally—my birthday! This year, I wanted to fly low, under the radar, no big deal—Joan Baez with some good friends on Sunday, lunch with a close friend, taking a blueberry pound cake I had made to my yoga group, and then having my young friend Rachael accompany me to meditation. All of this was sweet, and then, the best part. I noticed last night, as I inhaled the huge slice of watermelon one of my friends had given me, nothing anybody gave me was made of plastic or packaged in it. Wow! How lucky I am to have such thoughtful, generous friends who are paying attention to my quest. Cards, a good book, chocolate, fruit, and flowers from the farmers' market. And none of it needs to be thrown away. I did not plan this, but scout's honor, it happened.

It seems that life is full of surprises.

Thank you.

27. SUNLIGHT AND SHADOW

August 18, 2010

Not everything that counts can be counted, and not everything that can be counted counts.

—SAID TO BE A SIGN IN ALBERT EINSTEIN'S OFFICE IN PRINCETON, NEW JERSEY

For most of us, summer is over. People are returning from vacations all over the place. Teachers are back in schools, and if the students aren't, they will be soon. All of this and it is only mid-August, with a high of 95 degrees predicted today here in Durham. Monday morning, the online *New York Times* had a feature entitled "The Unplugged Challenge,"[29] where people all over the country volunteered to drop their media connections for a week. Writers began to write by hand, while kids gave their parents their cell phones for a day and stopped texting. Giving up Facebook was discussed again and again. As I listened to and watched the videos of these individuals, I kept thinking about how the internet, the Web, texting, etc. have become ubiquitous. I know a lovely ten-month-old with a Facebook page. How is all of this rapid knowledge and communication affecting us?

All of these streams of information have become integral to our daily lives. We pick, we choose, we answer our emails, we join Facebook or Twitter, or we don't. Always, we surf the Web expecting knowledge to be at our fingertips. Last week, I spent two nights at my cousin's house in the mountains near Swannanoa, North Carolina. I went mostly to visit friends and family and to spend time with her lovely new baby, but I did find myself alone for a brief moment in a mountain stream. The water was cold, rushing over the slippery rocks and roots of rhododendron. Solitude. A gift. A small diamond of time alone in the woods. A moment that I will keep in my pocket as I rush along the highways of my busy world. And yes, there was plastic in that remote stream—a black piece from the bottom of a PET soda bottle and a doorknob, both caught under stones and branches, surrounded by the flowing water.

Like cell phones, like the internet, like Facebook, plastic is something we will never live without again. For so many reasons, many of them medical, we would not want to. Somehow, even while tired and traveling, I have found that I can dodge the single-use stuff. This means not going through the drive-through, so that I can more easily ask for a drink without a plastic straw and plastic top. It means taking my steel water bottles along and keeping them filled. It means wishing I had remembered to bring some nuts and dried fruit along on my drive to Asheville, but even so, resisting the plas-

tic-wrapped candy at the gas station along the way. I never know if I am going to hold out, but somehow I do. I think I must really want this. To know it is possible gives me a sense of control, a tiny bit of power in a crazy world.

Here's some good news. Last week, my friend Rebecca (author of the blog *Less Is Enough*) and I got together, and she showed me how to make lotion! This stuff is amazing, and took only a short time to make. The recipe is below. Next, my cousin Monty (of granola fame) gave me a bar of handmade kitchen soap that a friend of hers made. I can rub this on a wet sponge and plenty of soap lathers up for me to wash the dishes in my sink. She has promised to buy me a bunch of these bars from her neighbor when she goes back to New Jersey this fall, plus try to get the recipe.

Meanwhile, enjoy these last few days of summer. If you live anywhere near Chapel Hill, don't miss Paperhand Puppets' *Islands Unknown* at UNC's stone-built amphitheater, playing now. Before the seasons change, I hope you are able to spend some time in a pool or lake, or maybe just look up at the moon and stars while the cicadas sing. This spinning world is always a place of mystery and amazement, when I remember to take the time to notice. Sunlight and shadow. It's all there.

RICH LOTION FOR CHAPPED HANDS

2 ½ ounces [total] apricot kernel oil and/or avocado oil
1 ½ ounces Shea butter
¼ ounce beeswax
4 ounces rose water
1 tablespoon pure vegetable glycerin
½ teaspoon grapefruit seed extract
20 or so drops of essential oil as desired

Combine the oils, Shea butter, and beeswax in a double boiler over medium heat, and cook until wax is melted. Remove from heat and add the rose water, glycerin, grapefruit seed extract, and essential oil. Blend with an electric mixer until creamy. Store in a glass jar.[30]

Tim Barkley

28. RESPONSIBILITY

September 8, 2010

Like all explorers, we are drawn to discover what's out there without knowing yet if we have the courage to face it.
—PEMA CHÖDRÖN, *When Things Fall Apart*

If you think you are too small to be effective, you have never been in bed with a mosquito.
—ATTRIBUTED TO BETTY REESE, American military officer and pilot

I gave an artist's talk last Wednesday at Guilford College during the opening for my installation *Again and Never Again: Can We Co-Exist with Ourselves?* I made the installation with help from more than 100 students, faculty, and community members. "Be casual," Kelsey McMillan, curator *pro tem* at the Guilford College Art Gallery, told me. I had lots of people to thank, and I did so, but not before I said something totally unplanned: "Artists are going to save the world." Did I really say that? I think so.

While I was installing this show, I told my story over and over again about my reluctant year without single-use plastic. Together, the students and I made a beautiful installation out of glass bottles, tin cans, bottle caps, and plastic beach detritus, among other things. These were smart students, interested and ready to help. These kids are anxious to make the world better, as best they can.

Afterward, Vernie Davis, a professor of Anthropology and Peace and Conflict Studies at Guilford, told me that the word *responsibility* means "the ability to respond." Right! Perhaps that is one of the very best things a good education can give us—the ability to see what needs to be done and the willingness to do it.

So many times this year, people have said to me, "I want to do this bring-my-own-bag thing, but I just forget." When I asked a very smart recycling educator whether paper or plastic bags were better for the environment, she responded, "Why on earth would you be letting a store give you a bag in the first place?"

This should not be a debatable point. Put that conversation together with the teenager who told me to go back to the car every time I forgot my bags, and I am trained. The students I worked with last week are ready to pay attention and ready for us to do so as well. What are we waiting for?

I told the students that they may think it's a hard thing, attempting not to use single-use plastic for the year. Hard, maybe, but really, it has been very interesting and invigorating. I will think something has stumped me, and then voilà, a solution appears. If I give myself a bit of time to look around or talk to a few people, then an answer arrives (like the used blender I needed that my friend Ann pulled out of her car's trunk).

People are doing great and thoughtful things, including a dedicated group of folks who spent the last weekend in August cleaning up an area of the New River. Taking action around and responsibility for the things we really care about is what life is about in so many intangible ways. Don't you

think? For information about how you, too, can help with such a cleanup, contact the National Committee for the New River (now called the New River Conservancy).

I can't end this post without mentioning Deep Roots Market, a small co-op in Greensboro started in 1976 at Guilford College. There, amid organic locally grown produce and milk in glass bottles, I found liquid dish detergent in a big vat. All I needed was my own glass bottle to fill up. Hallelujah! Who knew? Here's to the small enterprises in communities that give us choices and diversity in what we can buy. Here is what Deep Roots has to say about plastic bags:

> One of the most exciting and groundbreaking steps we (Deep Roots) have taken is to banish plastic bags from our checkout counters. Instead of giving out so many of these bags, which pollute our landscapes and seascapes, harm animals, and waste precious petroleum, we encourage shoppers to bring their own

> reusable bags. For beginners and those who forget, we still provide reused cardboard boxes from our shipments, when available, as well as paper bags (at a cost of 2 cents per bag, to cover our expenses).[31]

You might also want to watch a beautifully made short film, *The Majestic Plastic Bag—A Mockumentary*,[32] narrated by Jeremy Irons. He follows a plastic bag from the grocery store parking lot to the Pacific ocean's gyre of plastic. The film is only five minutes long, well done, and makes visible what we don't want to know about what happens to these bags.

I want to end this post by thanking the many students and faculty of Guilford College for their generosity, intelligence, trust, and time given to helping me with my installation. In no particular order, and with much appreciation, I thank Kesley McMillan, Theresa Hammond, Anthony Lowe, Mac McBee, Jim Dees, Vernie Davis, Kim Yarbray, Bob Williams, Adele Wayman, Chris Henry, Maia Dery, Kathryn Shields, Louis and Jerry Boothby, and Mark Dixon. The exhibition is in the Hege Library at Guilford College in Greensboro.

29. OPPORTUNITY

September 21, 2010

People usually consider walking on water or in thin air a miracle. But I think the real miracle is not to walk either on water or in thin air, but to walk on earth. Every day we are engaged in a miracle which we don't even recognize: a blue sky, white clouds, green leaves, the black, curious eyes of a child—our own two eyes. All is a miracle.

—THICH NHAT HANH, *The Miracle of Mindfulness*

Remember always that you are just a visitor here, a traveler passing through. Your stay is but short and the moment of your departure unknown.

—BHANTE DHAMMIKA, Australian Buddhist monk

Plastics made to last forever, designed to throw away.

—5 GYRES BLOG

It is mid-September in Durham. The other day, I had lunch under a wisteria vine with velvet seed pods that were waiting for a cool morning to pop open. Later, I went with two friends down to my favorite swimming hole on the Eno River and jumped right in. A bit of wind blew across the water, promising fall, but the water was warm and the swimming was easy. Also, I am finally putting the finishing touches on my old backyard shed, almost redone into a new studio!

I began this blog last January. which means I have been working on my year without single-use plastic for nine months now. The year is closer than ever to being over. Many people have been asking me what will come next. Will I be glad to end the year? Will I go back to plastic as usual, ignoring all of the packaging around stuff and buying it anyway? I don't think so—I just can't. I have written in other posts about the five gyres of plastic that are in our oceans, churning and swirling away, becoming non-food for wildlife, growing bigger and bigger and attracting pollutants. This past month, the Paperhand Puppet Intervention, in their show *Islands Unknown,* even made a large character out of a plastic gyre. At least the gyres are becoming common knowledge, I thought. When in conversation with friends and colleagues interested in my quest, I am shocked to find that many are not aware of the gyres. Going online after one such conversation, I found the *5 gyres* website (www.5gyres.org) with information and animation about all this. One of the first challenges the site gives is to go to the grocery store and try not to buy food encased in plastic, just to notice how much non-recyclable, only trash-able plastic there is around our food.

Amy Kellum, a friend and yoga instructor, sent me a link to last week's Huffington Post.[33] They wrote an article suggesting seven ways you might reduce your use of single-use plastic. It lists stuff like bringing your own bags and removing throw-away plastic bottles from your life. The article starts by suggesting that you cook your own food, something many people are learning to do for the first time. When I began this year, I had no idea so many others would be thinking like me.

This is the good news. The not-so-good news is that many of us are oblivious, and not interested in changing anything in our rush-ahead, convenience-driven American lives. As a society, we make things easy, with our plastic bags for everyone and free disposal of all our waste. I got an email a few weeks ago from a woman in Michigan who brought PAYT (Pay-As-You-

Throw) garbage collection to her town. Her town was running out of money and could no longer afford the alternative, which was throwing away vast amounts of waste, some of which would be recyclable. She told me that because of her stance of individual responsibility, she was verbally and viciously attacked as a communist, among other things. What part of promoting individual responsibility is communist? She wrote:

> The Solid Waste committee spent six months discussing my proposal. We had two packed public hearings during which the guy wearing a Rush Limbaugh T-shirt yelled, "Communist," and the old guy I beat for mayor led a group of angry seniors saying, "She'll use the savings for art programs," which was partly true. The irony of course is that PAYT is not communistic and the art program cost $1,000 of the $150,000 savings. After I left city hall, the garbage program remained but the art program was axed.

What will I do next year? I think I'll buy and eat a box of Honey Nut Cheerios, maybe a bunch of cheese, three or four boxes of good crackers . . . oh, and a bag or two of Ruffles potato chips. After that, I will be back at the farmers' market, back to not using single-use plastic. I have much more art to make about our overuse of it. What will it look like? How will it happen? I am not sure.

I do know that paying attention is very important, and my yearlong experiment confirms my agreement with jackson browne that single-use plastic is as much of a threat to us as global warming. Next year looms before me as a huge opportunity. We need to make changes in how we buy and consume. All of us. Some people will not do this until they are knocked in the head by the cost of our garbage or by the pollution of our oceans.

I say we continue to push for good environmental stewardship. Let's buy without plastic as much as we can and demand that the stores that sell us the products we consume also strive to minimize their use of plastic. Is this a pipe dream? Perhaps. Is this an opportunity? Absolutely.

30. SLOUCHING TOWARDS BETHLEHEM

October 13, 2010

> *The very least you can do in your life is to figure out what you hope for. And the most you can do is live inside that hope. Not admire it from a distance, but live right in it, under its roof.*
>
> —BARBARA KINGSOLVER, *Animal Dreams*

Some days I get frustrated and feel like I am on the dark side of the moon. Even though I work on being positive and look for the best of things, pumping myself up day by day to figure out again and again how to not use disposable plastic, I just get tired. I feel overwhelmed and wonder what I am doing. Why am I even doing this, when in every single store I see so much single-use plastic? What in the world am I doing?

Last Sunday, I spoke about my doubts with my friend Jenny on a walk along the Eno River. It was sunny. It was fall. We heard a few turtles plopping off their warm logs as we walked along the path closest to the water. She told me that I have made her look more closely at the world around her. She sees random trash on the ground, in the river, everywhere. She would have ignored it at one time. Now, when she can, she picks it up. The same goes with the plastic around food; now she notices, and she no longer allows anyone to give her a Styrofoam takeout container. Thank you, Jenny.

On Friday night, I had a small opening at Twig, which is a local store specializing in green products. A woman showed up because she was attempting not to use plastic and wanted to meet me. "Why?" I asked. "Why are you doing this?" Her answer had to do with reading about the gyres of plastic in our oceans.

Did this sound familiar? Yes, yes, yes! We talked techniques and details, and cheered each other on. Great!

The truth is, some days I do get tired of trying, of paying attention, of not getting to eat good takeout North Carolina barbecue because of all that Styrofoam. A few weeks ago, as I hiked the Detroit airport on my way home from Maine, I found I had left my steel water bottle in a bathroom along the way. I was tired, and I had lost my sense of humor and my perspective. Only the severe turbulence as we were flying toward Raleigh-Durham saved me from a plastic cup of juice served by the airline.

The overuse of plastic is so embedded in my consciousness these days that sometimes I am a zealot with no patience with others. I know this is not a good thing. We live in an amazing world, full of opportunities. A year ago, before I started *The Last Straw*, I used to love drinking out of straws, single-use petroleum product or not. It's a lot to figure out. Bit by bit, step by step, my hope is that we will get there, that we will figure it out.

Update: the PLA plastic (a.k.a corn plastic) cups that are supposed to be "biodegradable" are still whole and solid in my compost bin. I have written in earlier posts how PLA only biodegrades in big industrial composters and is a pollutant to other recyclable plastics. So this "answer" to reducing take-out garbage from our waste stream is only an answer sometimes. What kind of news is this to all of us in a hurry? Industries could fix this, but are not being held accountable by governmental regulations or by us.

It's a dance. How will I remember this year? I have learned that I can live a good life without most throw-away plastic. Unlike my friend Jenny, I have been noticing and dealing with the detritus around me for years. It is not going away. Are we as a society becoming more aware? Maybe. Do we need to? Yes. This year, I am on a journey, but I know that I am not one of the wise men. Perhaps I am a shepherd, not sure where the road will take me, certain only that the star I am following is hope. On this pilgrimage, the knowledge of my sense of humor, when I can find it, is what keeps hope alive.

Stephane Bidouze/Shutterstock.com

31. OMMMMM

October 18, 2010

There is no such thing as garbage, just useful stuff in the wrong place.
—ATTRIBUTED TO ALEX STEFFEN, American futurist

Have you seen this temple? It is the most beautiful, useful, sublime thing I have ever seen, and it is made out of once-used bottles.

A yoga teacher once told me that the word "Amen" comes from "Om." I'd like to think that's right. I am in awe of this Buddhist temple made out of beer bottles. I found it through a site called Tree Hugger. (Thanks, Janine, for telling me about this website!) It just doesn't get much better.

I am on the road this week and just arrived in Charleston, South Carolina, to make a mandala out of bottle caps in the rotunda of the College of Charleston's Addlestone Library. I have been invited, along with other amazing artists, to be a part of *bluesphere: earth art expo*. Tonight, I walked up George Street to King Street, scoping out places that serve coffee and take-out food without single-use plastic. I had great luck. This may be my biggest mandala ever. Gosh, that rotunda is **H-U-G-E**!

Om. . . .

Have a great week, all. I'll spend the rest of the week (October 19–22) at the library in Charleston making a mandala out of caps and lids. I have about 40,000 or so, plus what has been collected here. We could use your help. It will be fun. It will be beautiful. I hope lots of people come by to help.

Romolo Tavani/Shutterstock.com

32. BLUESPHERE

November 19, 2010

We all have very little time. So we must do everything as slowly as possible.
—Zen saying

Last Saturday morning, I was home and felt in my bones the privilege of being lazy. Outside, the sunlight glowed in yellow ribbons through the leaves, which are still on the trees. Since I last posted, I have been at work in Charleston, South Carolina, and Madison, Wisconsin, with a side trip to Chicago along the way. I have done a neighborhood art walk and moved into my old backyard shed, which has been transformed into my "new" studio. This studio is tiny, like a ship's galley. Each afternoon when I'm home, I try to find the time to leave my computer and my phone behind and make the trip down my back steps to the small building twenty feet from my house. I settle in and get lost in my work as a sculptor. These days, I am making small brown birds and other backyard wildlife.

So much has happened. I have traveled a lot of miles, on the road and in the air. I have met and talked with many people. I have seen old friends and made new ones.

This week, I went to Lexington, North Carolina, and next week I go back to Charleston to take down my bottle cap mandala. It was a part of *bluesphere: earth art expo* curated by Mark Sloan of the Halsey Institute of Contemporary Art. All this moving about, coming home in-between, has made me think more than ever about this great big "bluesphere" where we all live, hurrying about, eating and breathing, learning and looking. Always, the world is turning. In Durham, the summer is turning into fall, which brings yellow maple leaves glowing on the sidewalks and brown oak leaves filling my yard. Night and day, day and night.

During my travels, I have spoken with a lot of people about my mission to have a plastic-free year. People have been curious and, I think, inspired. I have learned so much about what we produce, use once, and then never use again. When I began the year, I had no idea that this journey would become such a passion for me. Sometimes when I hear myself talking, I sound shrill, a little over the top, way left of center. "Okay," I say to myself, "Let's take a deep breath here. Slow down . . ."

When I was in South Carolina working on my mandala, I met the inspired artist Chris Jordan. Among other things, he has chronicled our use of stuff and its quantity in a very powerful series, *Running the Numbers.*[34] A few weeks ago, I called Muriel Williman, the Education Coordinator for Orange County, North Carolina, Solid Waste. I wanted to check my statistics before I gave an artist talk. "We are at the beginning of our awareness of this plastic thing," she told me. At the beginning of our awareness of the magnitude of our overuse of plastics.

About a month ago, a friend from Switzerland sent me a link to Dianna Cohen, a West Coast artist, speaking "tough truths about plastic pollution."[35] Simply and clearly, she urges us to pay attention and tells us how she has reduced plastic in her own life. Among other things, she urges us to add a fourth "R" to the Reduce, Reuse, Recycle mantra—*Refuse*. I encourage you to look for her work.

Because of NASA, we can see the world as a ball, floating in space. A tiny speck in the vast space in which it spins. Have you ever held a live tadpole or a small fish in your hands so you could see its heartbeat, the coursing of blood through its veins? A live thing, held together by a thin, translucent layer of skin, bone, and muscle. To me, that is what earth looks like from so far away. A live ball, pulsing and breathing, with all the oceans and rivers and species living under its skin.

Today, I listened to Salman Rushdie talking about his new book, *Luka and the Fire of Life*[36] on NPR's Diane Rehm Show. He said that here on earth, we humans are the only species that tell stories to each other. We read books. We watch movies. We tell each other the things we feel are most important.

I have heard one story more than once this month from artists Chris Jordan and Dianna Cohen. There is so much plastic floating in our oceans now that we cannot possibly clean it up. At least, not until we stop the constant flow of it from the rivers to the sea. The plastics we use once are flowing into the oceans at an alarming rate. Birds and other sea creatures are dying. As humans, we need to tell this story. The animals cannot.

33. WHAT DO WE WANT?

December 16, 2010

The events in our lives happen in a sequence in time, but in their significance to ourselves they find their own order . . . the continuous thread of revelation.

—EUDORA WELTY, *One Writer's Beginnings*

Write about what you don't know about what you know.

—GRACE PALEY

Convenience might kill us in the long run. I grew up hearing the frantic warning, "Don't ever put that plastic dry cleaning bag over your head; you could smother yourself!" Now, with equal alarm, I think about what single-use plastic is doing to our world.

For the past few days, I have been listening and re-listening to a Radiolab podcast of a conversation Robert Krulwich had with Steven Johnson and Kevin Kelly. The show was called "What Does Technology Want?"[37] Johnson and Kelly are authors of *Where Good Ideas Come From*[38] and *What Technology Wants.*[39] Think about it this way—stuff gets invented because the world needs it, and we want it. Because our world is ready and waiting for it. For example, Elisha Gray registered his invention of the telephone three hours after Alexander Graham Bell did. The telephone's time had come; it was waiting to be invented. The same thing happened with the electric light bulb. The scientific world had already made all the foundational discoveries necessary to support these inventions. According to Johnson and Kelly, there are no real "Eureka" moments. Stuff happens because of all that has come before it. They also say that ideas, like plants to the sun, lean in the direction that the world wants.

Are we ready to address our overuse of plastic? Do we want to pay attention to the fact that we are using oil, a finite resource, in a nonrenewable way? For that matter, are we ever going to make small, affordable cars that use oil wisely? What must happen scientifically and culturally to make conservation a priority for us?

If you listen to NPR, you hear stories about electric cars and wind-generated energy, which makes me think we are leaning toward some sort of environmental stewardship. I do not hear much these days about plastic, how it is used in almost all packaging, from food markets in the developing world to our fancy gourmet cheese counters here in the States. So much of our food is sealed in plastic for freshness and shelf-life, yet no one is talking about the toxic off-gassing of the non-biodegradable materials we are using to hold our food. Plastic sheeting, plastic wrapping, plastic clam shells, plastic lids with plastic straws covering plastic drinking cups.

I am finding that I have more questions than answers as I approach the end of my year without single-use plastic. I feel I have been mostly successful in my quest, saying, "This is my job this year," with a sense of humor, plus a bit of vigilance. I have found most of the food I need available in glass or at farmers' markets. Everywhere, however, on almost everything, are tiny plastic seals, tabs, labels, and caps—*every* everywhere. I also see, all the time, and everywhere as well, people filling their trunks full of groceries in plastic bags, folks strolling down the streets or sitting in internet cafes or waiting for buses with their giant "big-gulp" drinks of something, usually in a plastic cup and almost always with a plastic straw and lid. Plastic water bottles are everywhere—along our roads, in offices and homes and cars and in people's hands. *Everywhere*. Always, always, always. I ALSO see kids and adults with their own use-again water bottles and lots of people with their use-again bags at the grocery store. It's all there. My questions: Are we leaning toward paying attention? Do enough of us feel that the time has come to do so? Steven Johnson says if an idea comes before it is supported by the ethos of the community, then it will flounder. I hope we are ready to deal with this plastics issue. I have been wondering a lot about what it would take for us—our society, our culture, our world—to begin to take the over-use of plastic seriously. What would make us shift our awareness and our actions? It is becoming clear to me that looking for the answers to these questions will be the next stage in my journey.

34. AROUND THE WORLD

December 24, 2010

Without playing with fantasy no creative work has ever yet come to birth. The debt we owe to the play of imagination is incalculable.

—CARL JUNG, *Psychological Types*

To raise new questions, new possibilities, to regard old problems from a new angle, requires creative imagination and marks real advance in science.

—ALBERT EINSTEIN, *The Evolution of Physics*

Around the world, people are making Christmas trees out of discarded stuff. Everywhere. I found lots of Christmas trees on the Web this morning made out of things as varied as bicycles, wooden pallets, plastic silverware, and Mountain Dew cans. Here's one I made for the local arts center out of glass bottles, tin cans, and plastic refuse from the beach (see above).

It seems everyone is doing it, everywhere. Recycling creatively and having a good time with it. So, here's to new ideas, new possibilities for us in recycling and reusing all we make and use on this planet earth.

The world is full of creative minds. Like bees, we are humming along. Perhaps this season we are even singing. Merry Christmas, everyone. Thanks to all of you who have helped and supported me in my (no longer reluctant) year without single-use plastic. This has been a full year for me. I have learned much and see much more learning and understanding, thinking and doing along the road ahead. As much as anything, my inspiration this year has come from you, the readers who have questioned me, educated me, and inspired me. I thank you all the time.

35. TO DO WHAT IS RIGHT, NOT WHAT YOU SHOULD

January 19, 2011

No problem can be solved by the same consciousness that created it.

—ATTRIBUTED TO ALBERT EINSTEIN

Power without love is reckless and abusive, and love without power is sentimental and anemic. Power at its best is love implementing the demands of justice, and justice at its best is power correcting everything that stands against love.

—MARTIN LUTHER KING JR,
Where Do We Go from Here: Chaos or Community?

Floundering? Swimming upstream? Just plain wondering what I am doing? After this past year, striving in every possible way to use NO disposable plastic, I find that my eyes and heart are so accustomed to dodging this stuff that I cannot stop. Or, put another way, I guess I do not want to.

A friend just sent me the Blue Hill, Maine, co-op newsletter. The community had just watched the documentary *Bag It.*[40] Here is a startling fact from the movie I learned through the newsletter: "We have created and used more single-use plastic in the last ten years than we did in the entire 20th century." Bag it, bottle it, seal it, cap it, cover it, keep it separate, keep it clean, drink it up, toss it away. It seems we have just begun our consumption of single-use plastic here.

"To do what is right, not what you should." This is the message a friend of mine received in her very first fortune cookie of 2011. As I sort through the memories of the past year and all the plastic stuff I did not use, but needed to, stuff like glue and tape, and stuff I did not use but really wanted to, stuff like store-bought crackers, which all come wrapped in plastic, I am wondering which of these things I am going to allow back into my life. Most stuff—take-out plastic, plastic bags, straws, Styrofoam of any sort—I am happy to continue not using.

I have been listening to a collection of tapes called *The Monticello Dialogues*[41] by William McDonough. These tapes were lent to me by a young man behind the cheese counter at my local Whole Foods who has been wrapping my cheese in paper this past year. Like Thomas Jefferson and Buckminster Fuller, McDonough is a visionary. He believes we can figure this out and make a better world.

McDonough reminds us of our "timeful mindlessness." We are all in a hurry, don't have time to think about our choices. He suggests "mindful timelessness" instead. In short, I think he is saying, give yourself the space and time in your life to do the things you can to care for the earth—recycle, bring your own bags—whatever this is for you. And he encourages you to see this earth as a place of vast abundance, where a better and more balanced world is possible if we put our minds and hearts into it.

And by the way, did you hear that the country of Italy has banned the single-use plastic bag? Yes, I said the *country* of Italy!

These past few weeks, as I work in my backyard studio and see the birds—Carolina wrens, cardinals, nuthatches, and more—stuffing themselves with the seeds I am putting out, as I see the daylight lasting longer each afternoon, as the tips of the daffodils show themselves under the damp humus and melting snow, I know that spring is indeed on the way. The earth has shifted and spring with its awakening is coming again. So, late in this January 2011, I wish you all a Happy New Year. I have changed the name of this blog to *The Last Straw: A Continued Quest for a Life without Single-Use Plastic.* Certainly I have not found perfection in this quest, but I am still curious and determined. The question of how we will use the resources found in the earth during this new century is a big one. How will we provide for all of us all over the world? Recently, I have been attending some neighborhood meetings sponsored by Clean Energy Durham, where we are learning to make our homes more energy efficient in simple, direct, pass-along ways. This organization encourages us to learn and then share our knowledge with our neighbors. It is a good model for community and for environmental awareness and empowerment. I told my young hosts, Chris and Irene, about my commitment to use as little disposable plastic as possible. Chris said, "You know, I have discovered that if I actually drive at the speed limit, then I save four miles per gallon." Wow. It is these individual acts of paying attention that inspire me and keep me curious. What will the new year bring? For myself, I am hoping to figure out what is right for me, so that I will not act righteously (I *should* do this and *should* do that), but act from my core. I hope to do the best I can to keep figuring out what I want to do, and doing it.

PART TWO

THE FOLLOWING YEARS

36. BREATHING

. . . .

February 11, 2011

I look for what needs to be done. After all, that's how the universe designs itself.
—BUCKMINSTER FULLER

Never underestimate the power of compassionately recognizing what's going on.
—PEMA CHÖDRÖN, *The Places That Scare You*

Last night at a meditation group, my teacher suggested that when we wake up in the morning, we pay attention to our first breath. Is it an inhale or an exhale? From there, she went on to suggest that we allow our day to unfold, being mindful both as we go along and as we remember. Breathing. Just in case you are curious, I awoke on the exhale this morning.

The day is cool and sunny. On the way back from swimming at the Y, I remembered to think about my breath again in the car. I thought to myself, "All this is interesting, but what does breath have to do with my daily life?" And then an inner voice answered me, quietly and very definitely, "Everything, Bryant, just everything."

After my last post, which was also my first post of the new year, friend and fellow blogger Rebecca Currie asked me a big question: "So, Bryant," she said. "How did you do it?"

Rebecca, here are the answers, as best as I can figure:

1. I "did it" because I really wanted to. I saw it as my job. I was vigilant. I said no to the plastic lids on cups, to plastic straws. No to plastic bottles in stores. Many bottles look like glass, but when you feel them, they are plastic. No to all Styrofoam takeout containers. No to plastic bags in stores, so easily given to us. Lots and lots of "No, thank-you's."
2. I went to my local farmers' market almost every Saturday. I found un-bagged produce that I could slip in my own reusable bags. Bread, eggs, and baked goods were always packaged in paper.
3. I brought my own glass bottles and metal tins to Whole Foods, and they weighed them for me. Then, I could get things from the bulk bins. Where I live, we also have Weaver Street Market, a co-op that also has bulk bins. I would not have been as successful as I have been without these alternatives.
4. Cousin Monty's granola, full of almonds, flax seed, and many other good things. Making this every three or so weeks meant that whenever I was hungry and needed a boost, it was there and it was good. Really good.
5. I made my own yogurt. This proved to be much easier than I had ever imagined and very good as well.
6. I carried my own steel water bottle. To make this work, I kept three in circulation in case I forgot one, which was often.

I am in a groove. No stopping now!

As an experiment, and to give myself a break, I bought a few things packaged in plastic this month: a box of Cheerios, two Snickers bars, and two boxes of crackers. I enjoyed all of them, especially the Snickers bars. High fructose sweet sticky stuff is a powerful drug. I wanted to see what I was missing. What I really want to continue buying is, believe it or not, crackers. Even this may change, as my young friend Lydia has sent a group of passionate emails about her new-found love of making flat breads like chapatis and Korean pancakes. She says they are amazing. (A note on packaged crackers here—the stuff inside the cardboard box is not waxed paper. It is made to look like it, but it is not. Try putting it in your compost, and you will see that it does not decompose; yep, it's plastic.)

Recently, several people have forwarded a video to me of news commentator Van Jones speaking of plastic pollution and its connection to economic injustice. He makes a powerful case for paying closer attention to how we use (and misuse) plastics in this world.

Jones also speaks of our addiction to disposability, how we feel good about putting our plastic bottle in the little blue bin but do not consider the consequences of recycling it (i.e. burning it, thus releasing toxic fumes into the environment). This is happening more frequently in Asia, where environmental air standards are much lower than those in the US. Plastics burning in Asia have wiped out clean air gains in Los Angeles, which are now back to their pre-1970's levels. Jones challenges us to think about the very idea of disposability—of species, of raw materials, of people themselves—because it is the people who work to make and "recycle" plastic products who suffer the most.

The very air we breathe is suffering from our addiction to a "use it once and then toss it" mentality. And all of us are breathing—every diatom, every cell, every species. Inhale. Exhale. Repeat and repeat again. To live, to be alive, we breathe. Spiders, earthworms, palm trees, porcupines, whales, goldfish, monkeys, dogs, and people. All species, everything alive on earth.

The breeze at dawn has secrets to tell you.
Don't go back to sleep.[42]—Rumi

37. LETTER TO MICHAEL

March 23, 2011

We are called to be architects of the future, not its victims.

—BUCKMINSTER FULLER, *Critical Path*

We are continually faced by great opportunities brilliantly disguised as insoluble problems.

—ATTRIBUTED TO LEE IACOCCA, CEO, Chrysler Corporation

For the past month I have been traveling, mostly in North Carolina, doing residencies in schools. Though I love this work, I miss home and am trying to figure out where to eat in strange lands. If I am very lucky, I stay with friends, but more often I am on my own. While at a terrific residency at Rosewood Elementary School in Wayne County, North Carolina, K&W Cafeteria became my home away from home—a place of vegetables served on washable plates and no plastic utensils.

Weekends have been time for laundry, with fleeting glimpses of friends, a basketball game or two, plus hunting down the trout lilies that are blooming along the Eno River.

For the past weeks, I have been listening to stories of the tsunami in Japan, feeling horrified by the destruction, sorrow for the many deaths, and a frustrated feeling of wanting to do something. A 9.0 earthquake shifted our earth's axis and the tsunami came after it, both out of our control. We feel powerless in the wake of such devastation. The tectonic plates of our very earth shifted.

During a three-hour drive to the Atlantic Ocean, with the sun shining and green buds on trees, I felt both distanced from this disaster and overwhelmed by it. What can I do here? Nothing? About the tsunami, probably not. However, I am reminded again and again that it is the things that I *can* do, however small, that are important. For me, in the whirl of travel and residencies, I have been putting compost on my garden as I watch the tree limbs change color. They go from stark gray to muted shades of brown, green, orange, and red as their leaves begin to unfurl. I have been dodging single-use plastic, seeing that I can do it every day. In stores, there are usually options, and these days with all the new food trucks around, I can even find takeout food in brown paper bags!

For you, the things you can do might include riding your bike somewhere instead of driving, or stopping to show a young child in your life the flowers of spring as they unfold. Walk forward and, as one of my meditation teachers has said, soften yourselves into the things you must do. Keep active, move along, in delight and joy when it is easy, and with courage and kindness when it is not.

Below, I am posting Buckminster Fuller's letter to Michael, a ten-year-old boy who wrote to him in 1970. You can find it in his book *Critical Path,* but I have also seen it reprinted many times. When I am teaching, I often read it at the end of a class. From my early twenties as a young artist, I have kept this letter inside me like a heartbeat.

> Dear Michael,
> Thank you very much for your recent letter concerning "thinkers and doers."
>
> The things to do are: the things that need doing: that *you* see need to be done, and no one else seems to see need to be done. Then you will conceive your own way of doing that which needs to be done—

that no one else has told you to do or how to do it. This will bring out the real you that often gets buried inside a character that has acquired a superficial array of behaviors induced or imposed by others on the individual.

Try making experiments of anything you conceive and are intensely interested in. Don't be disappointed if something doesn't work. That is what you want to know—the truth about everything—and then the truth about combinations of things. Some combinations have such logic and integrity that they can work coherently despite non-working elements embraced by their system.

Whenever you come to a word with which you are not familiar, find it in the dictionary and write a sentence which uses that new word. Words are tools—and once you have learned how to use a tool you will never forget it. Just looking for the meaning of the word is not enough. If your vocabulary is comprehensive, you can comprehend both fine and large patterns of experience.

You have what is most important in life—initiative. Because of it, you wrote to me. I am answering to the best of my capability. You will find the world responding to your earnest initiative.

Sincerely yours,

BUCKMINSTER FULLER[43]

I am ready to do what Mr. Fuller said, or so I hope.
Meanwhile, here's to mud between your toes and spring beauty in bloom.

38. THE RED WHEELBARROW

April 26, 2011

When you blame others, you give up your power to change.
—ATTRIBUTED TO DOUGLAS ADAMS

You know, I think if people stay somewhere long enough—even white people—the spirits will begin to speak to them. It's the power of the spirits coming up from the land. The spirits and the old powers aren't lost, they just need people to be around long enough and the spirits will begin to influence them.
—CROW ELDER

It is the "before the bugs" time of spring in Durham. The leaves are full and green on the trees, and lawns have been mowed for the first time this year. The rains have kept the pollen down. My neighbors and I were lucky with the recent storms; just one tree down, fallen into the street. Now, with one fewer large oak tree in the neighborhood and much more sun reaching the ground, I am thinking of planting tomatoes for the first time in years.

Meanwhile, I have been working in schools throughout my region and even spent last week at the coast. I watched the moon rise most nights, an orange ball coming up on the eastern rim of the Atlantic Ocean. It has been good, this morning, to spend some time working in my yard with my new red wheelbarrow.

My work monitoring single-use plastic continues. Several weeks ago, I had the great pleasure of working with the Marine Ecology class at Fuquay-Varina High School, making the ocean food web out of discarded single-use plastic. First, we cleaned the creek next to the school of plastic detritus that had blown over the fence next to the parking lot. The students also brought in throwaway plastic they had used in their homes that week. Students selected animals from the ocean food web, from whales at the top to plankton at the bottom, and made them out of this collected plastic. We hung the food web, made entirely of cast-off plastic, for their Science Night. The students entitled it *Killer Plastic*. Among many wonderful creatures, you could see an ethereal jellyfish made out of plastic bags (floating top left).

Last week, I got to stay in a small motel right across the road from the Atlantic Ocean while I did a residency with the students of Grandy Primary School in Camden, North Carolina. The weather was warm and, every day after work, I took my green folding chair down to the beach. Each afternoon, I walked over the small dune that separated the ocean from the road and read my book, watched the birds and surfers, dug my feet in the sand, and in general felt very lucky. It was a short walk, and each day it was the same walk, yet the continually arriving plastic I found along the beach and on the path through the dunes was plentiful. It was not hard to find. I just picked it up as I walked back and forth. The beach is big; it looks empty of this stuff, but it is not. I continually found plastic bottles and plastic bags, both the kind from grocery stores and the clear ones around snacks. Then there are the bottle caps, old faded balloons, and bits and pieces of other things. More things that our marine wildlife might swallow. A lot of it gets thrown in the trash can beside the pathway over the dunes. Anything that doesn't make it into the can gets swallowed by the ocean pretty quickly.

What happens to the stuff that ends up in the ocean? Again, the ocean is very big. The health of marine wildlife, such as sea turtles, is always in my mind and is one of the reasons I pick up loose plastic when I see it floating around or half-buried in the sand. I am not the only one who does this.

Meanwhile, I live my life. Some days are easier than others. My favorite times are always simple ones. Today, I noticed how green my yard is, how damp and ready to grow things the soil has become. On a whim, I bought the red wheelbarrow pictured at the top of this chapter a few weeks ago at a small hardware store out in the country. Did I buy it to care for my garden or for the nurturance of my soul? I would say it is a toss-up. With this simple machine, a wheel attached to an inclined plane, I move compost to my garden, leaves to the compost, dirt here and there. Working in my yard nurtures me. Turning my compost, or digging in the garden where I turn the soil and find earthworms galore, I am happy. In my yard, I shift dirt about, weed and plant, plant and weed, listen to the birds sing, and wait for the hydrangeas to bloom. The world is big, and we are all busy in it. Worms and a wheelbarrow bring me home. There are no chickens in my yard, but I do see an occasional white-tailed rabbit, happy in the grass.

The Red Wheelbarrow

so much depends
upon

a red wheel
barrow

glazed with rain
water

beside the white
chickens

—William Carlos Williams

Courtesy of the family of Clara (Kitty) Couch

39. BAG LADY

May 18, 2011

Do the difficult things while they are easy and do the great things while they are small. A journey of a thousand miles began with a single step.

—LAO TZU

North Carolina potter Clara (Kitty) Couch was way ahead of her time, recycling-wise. While most of us were just beginning to understand the importance of curbside recycling, Kitty got the full picture of "reduce, reuse, recycle." She took her own bags to the grocery store long before it was a common activity. It has taken me many years to act on this concept, and I love that Kitty was out there early on doing the "remember to reuse" thing! Plus, as you can see from the newspaper photo, she was clearly having a good time with it!

In 1989, Kitty and I were part of a group of nine women selected to represent the state of North Carolina at the National Museum of Women in the

Arts in Washington, DC. It was then that I began to love her amazing ceramic works. They seemed to be alive, large, unglazed, almost abstract, always evocative. Not to mention Kitty herself, who was direct and joyful, her eyes wide open, attuned to the world around her.

I have just returned from an event in the North Carolina mountains, full of artists of all sorts from many parts of the world. I was thinking of Kitty a lot; a while back, at an earlier iteration of the event, I went with her to visit her home in the mountains. One of the things I remember most about that day was Kitty telling me about her firewood. It was stacked at the bottom of her road, and every day as she took her morning walk or went out to check the mail, she brought back a log. Bit by bit, piece by piece, she moved a stack of firewood up the hill to her home to be burned in the cold of winter. It was a ritual. She took what could have been a hard job and gave it a simple and everyday joy. I can imagine her looking at each piece of wood as she carried it up the hill. She had made this task, easily done in one quick dump by a guy with a truck, into a daily ritual.

Going back to Black Mountain where I had last visited with Kitty, it was natural to remember her. Meeting her friend Pinky helped as well. I became aware of how the story that Kitty told me twenty years ago about her firewood has become a part of my life. Each day as I move from room to room, I usually take something that belongs where I am going. When I weed the garden, I do it in bits and pieces, on the way to the trash can or out to the car. Bit by bit, I care for my world. Step by step, I shift and shape, and always I think of Kitty moving her wood pile from the road to the house, piece by piece.

In a way, life is a string of small, seemingly unimportant tasks. Ritual. Continuity. Repetition. Continuity again. We must have faith that our small actions count. We must believe in them. And if we are fortunate, we discover ways to find joy in these day-to-day tasks. The small acts of our lives are as important to us as leaves are to a tree. And all of them, whether we are carrying firewood, caring for loved ones, or sorting our socks, add up to the tree that is our life. Bit by bit, we become who we are. Everything we do counts. All of it. Remembering Kitty, I am pretty sure this is true.

40. BE STRAW-FREE

July 8, 2011

This morning, I counted twenty-one tomatoes on my four tomato plants. The peaches on my tree that sprouted from my compost pile will be ripe soon, and it has been raining every night here in Durham for the past three or four days. These are all good things. Summer is here and "the cotton is high," or at least promises to get there if the rain keeps up.

Here are two new initiatives that have excited me:

1. A nine-year-old boy named Milo Cress from Vermont has started a "Be Straw Free" campaign in his state. His website tells us that each day we use 500 million straws—enough disposable straws to fill over 46,400 large school buses per year. He is asking restaurants in his home town of Burlington, Vermont, to sign a pledge

to give people a choice whether or not to use a straw. People are listening; for one, the governor of Vermont. You can go to Milo's website and find out more about what he is doing and sign your own pledge if you so choose.[44] This young man has research behind him and a plan in front of him.

2. A grocery store that uses no packaging is planning to open in Austin, Texas. The name of the store is "in.gredients." It promises to be "a collaborative effort between business, community, and consumers with the goal of eliminating food-related waste while supporting local businesses and farmers." Go in.gredients! I wish you great success in your new enterprise.

Meanwhile, I continue to feel fortunate to be able to go to my local Durham Farmers' Market, twice a week if I want. I can get fresh local produce and bring my own packaging. Some new neighbors have offered me access to their front yard, which is full of ripe tomatoes and cucumbers as I write this, and the blueberries are ripe for the picking all around.

All of this is good news. The hard stuff is still how much non-recyclable single-use plastic I find around me. After more than a year working on being single-use-plastics-free, sometimes I still feel overwhelmed.

Right now, I want to say thanks to everyone who has sent me information and everyone who is changing their lives. My friends have told me about some of the things they have done since I began my quest to use less plastic in my life. They have bought refillable coffee mugs, begun to compost, begun bringing their own bags to the grocery store, and stopped using single-use plastic water bottles. I feel grateful for the attention these friends are paying and also grateful to the many people I meet from day to day doing the same.

Plastics are in our lives to stay. We love our computers and our shoes and our drainpipes and our swimsuits, our tennis rackets, our cell phones and our plastic tubing, fountain pens, lawn chairs, and flyswatters. Many people are working on making the industry more sustainable. More and more plastics are becoming recyclable, yet many still are not. All of this paying attention can be hard work. It can also feel good. I mean, my compost pile is amazing, and a very active place.

Lately, I have been struggling with my own righteous indignation over the glut of plastics in our lives. Why aren't more people bringing their own bags or not using straws? Why isn't recycling easier? Why is the plastics industry keeping "recycle" arrows around numbers on disposable plastics which are not recyclable? Why? Why? Why? I do not have answers yet as to how to deal with all I have been feeling. I know I feel a responsibility to keep learning about our earth and how plastic is affecting us. Also, equally important, are initiatives people are taking to heal the earth.

41. WILD GEESE

January 2, 2012

WILD GEESE

Meanwhile the wild geese, high in the clean blue air,
are heading home again.
Whoever you are, no matter how lonely,
the world offers itself to your imagination. . . .

—MARY OLIVER, *Dream Work*

The geese are traveling up above us this time of year. I saw a perfect V of them Friday night when I was out in the yard with Adele Rose, the soon-to-be two year old in my life. The sun had set, the sky had a just-before-dark, near-purple hue to it. The birds flew in close to us, almost right above the house. We could see their great flapping wings and long necks. Adele Rose, who loves owls, was mightily impressed.

Tonight, the weekend over and the work week soon upon me, on a whim I picked up a book of Mary Oliver's poems. I read a few of them until finally I settled on "Wild Geese," an old friend. It's funny how you think you know a poem, that it is inside of you already . . . and then when you read it again after a long absence, it is a new thing altogether.

I think what happens is life itself moving over our bodies and our souls day in and day out like a river, swirling about us. In its current, we swim, we breathe, we live, we laugh, we grieve, we worry, we grow, and change. Life does not stop for us, not one bit.

"Geese," I said. "Geese," Adele Rose answered back.

Happy New Year, everyone!

42. *KAIROS* AND THE WAY OF THE WARRIOR

February 11, 2012

> *Rejoicing in ordinary things is not sentimental or trite. It actually takes guts. Each time we drop our complaints and allow everyday good fortune to inspire us, we enter the warrior's world.*
>
> —PEMA CHÖDRÖN, *The Places That Scare You*

Recently, I have been feeling that my life is out of my control. More specifically, whatever I do, no matter how closely I do or do not pay attention to the details of my life, I am not in control. Stuff happens. A friend's brother gets a cancer diagnosis. I lose my wallet. I forget the most important thing I needed at the grocery store. Some things I might have prevented by being more mindful; losing and forgetting are certainly reminders to slow down. Other things, like the serious illnesses of those we love, bring us deeply into the vulnerable places in our souls. In the middle of all this, we work and love and get our bills paid as best we can.

Before Christmas, while in a bookstore looking at the staff picks, I found a wonderful book entitled *Crow Planet: Essential Wisdom from the Urban Wilderness* by Lyanda Lynn Haupt. I have been watching and making crows for over four years now, so this book jumped into my hands. Crows, and the way they are adapting to our rapidly changing world, are harbingers of the environmental changes around us. Many of us have crow stories. In fact, keep your ears open the next time you are outdoors; wherever you are, if you listen closely, chances are you might hear their call.

Haupt says the following in her introductory chapter, "Crows and *Kairos*:"

> There are two Greek words for time. One is *chronos,* which refers to the usual, quantifiable sequential version of time by which we monitor and measure our days. The other word is *kairos*, which denotes an unusual period in human history when eternal time breaks in upon chronological time. *Kairos* is "the appointed time," an opportune moment, even a time of crisis, that creates an opportunity for, and in fact demands, a human response. It is a time brimming with meaning, a time more potent than "normal" time. We live in such a time now, when our collective actions over the next several years will decide whether earthly life will continue its descent into ecological ruin and death or flourish in beauty and diversity.[45]

I do not want the part of me that feels "out of control"—either in my daily life, or in the face of the environmental distress I see around me—to be the part that takes over my world. Where I live, the daffodils are blooming, maybe early, but abloom they are, and they are gorgeous. Their yellow heads are bobbing at the edges of parking lots where I see plastic trash in the form of to-go lids, straws, and candy wrappers, abandoned and ignored.

To be a warrior as Pema Chödrön describes, to love the beauty I see around me, is a gift of living fully. It is a job I set my mind to daily, to balance my joy in the ordinary beauty of everyday life with the environmental waste and denial I see around me. I pick up what I can and make art out of it. I continue to use as little single-use plastic as I possibly can.

Ultimately, this is what I know. It is not about what I do, but what *we* do, as a society and as the people of this earth. Some days I am a warrior, and some days I am overwhelmed by the monumental task of paying attention and knowing that so much more needs to be done.

43. GOING UP THE MOUNTAIN

March 27, 2012

Sometimes the questions are complicated and the answers are simple.

—ATTRIBUTED TO DR. SEUSS

Tomorrow, I travel to Appalachian State University in Boone, North Carolina, in order to begin my residency at the Turchin Center for the Visual Arts. With the help of staff and students, I will make an installation entitled *STUFF: Where Does It Come From, and Where Does It Go?* The Turchin Center was a church before it became an arts center, and it is a lovely building. The exhibition space has high ceilings and large windows. The staff members have been collecting lots of plastic water bottles to fill the windows. I am bringing my ubiquitous collection of plastic lids and straws, which I have been picking up this year. I am fortunate to have been given the opportunity to work in such a beautiful arts center, and I know it will be a privilege to work with the students as well.

Last week, I helped the third and fourth grade students of Chocowinity Primary School in eastern North Carolina make a coral reef out of plastic bottles. We used lots and lots of PET bottles—mostly water bottles, some soda bottles, too—and made a very beautiful installation. The students made schools of plastic fish and lots of plastic seaweed. We had lots of fun using all those single-use plastic bottles to transform a breezeway in the school into an underwater grotto.

Where do all these bottles come from? Do any of you remember the advent of curbside recycling? I do. In my hometown of Durham, it happened over twenty years ago. I also remember that in the beginning, recycling plastic was controversial. Our trucks did not pick it up, because there were no markets for it. Plastic did not recycle full-circle; rather it would be down-cycled into other objects, but not recycled into itself multiple times. Many more things in those days were in glass containers. By now, plastic containers are such a part of our world that most people assume we could never live without them.

Plastic is recyclable, but not biodegradable. It comes out of the earth, but unlike leaves and rain, it does not return to replenish the cycle of life. It's possible that our point of view might change with the rising cost of petroleum. Plastic and gasoline mostly come from the same source. Petroleum is an extremely valuable and non-renewable resource. Plastic, made to use once in order to fit our single-use lifestyle, does not return the value it may at first seem to have. You have heard me say all of this before, and it is still true.

My friend Sarah B., one of the best collectors of stuff in my life, sent me an interesting piece of news. A bunch of colleges in the Northeast have begun to ban plastic bottles from their campuses.

44. STUFF: WHERE DOES IT COME FROM, AND WHERE DOES IT GO?

April 11, 2012

In the United States we buy stuff all the time, and very often we are not responsible for where it goes when we are finished with it. I look in trash cans frequently. In a way, as an environmental artist, I consider it my job. What I know is that, just by looking, I see lots of aluminum cans and plastic bottles, tossed rather than recycled.

I spoke with a woman from Texas today. She is a curator in a museum in Odessa, Texas, where prairie dogs are her neighbors. She told me they have a big problem with plastic bags polluting the environment; apparently, the bags get stuck all over the tumbleweeds. This must be unsightly, not to mention bad for the airborne plants.

Yesterday, while pumping gas, I chased, but did not catch, a yellow Walmart bag as the wind yanked it into the air and then threw it into the middle of a busy intersection. With lots of cars rushing past, I gave up my mission and stood, watching as the bag filled with air, twisted and turned down the road, on a journey to the ocean, perhaps getting stuck for a while in a tree or a tumbleweed along the way.

For the past two weeks, I have had the good fortune to be an artist-in-residence at Appalachian State University in Boone, North Carolina, at the Turchin Center for the Visual Arts, installing my show *STUFF: Where Does It Come From, and Where Does It Go?* I worked with Ben Wesemann and his intrepid cadre of students from the Catherine J. Smith Gallery, as well as the staff at the Turchin Center. By the time my residency was over, we had worked with over 200 staff and students installing 10,000 PET plastic bottles in the windows, making a comet out of bottle caps and beach plastic, and erecting a bamboo forest of chopsticks and Mountain Dew bottles. It was an amazing and humbling experience. Appalachian State recycling coordinator Jen Maxwell guesstimates that around 10,000 plastic bottles are

recycled each week at the university. Helping to clean and stack 10,000 bottles made me realize in a visceral way how much plastic that is.

My statement for the show:

> I saw it on *60 Minutes* tonight, so I know it must be true. Our oceans are filling with bits and pieces of plastic. The plastic comes from us and arrives in the ocean via our rivers, our streams, the wind, and roadways. It is going there all the time, and the fish and birds are eating it. Plastic is polluting the waters of our oceans in big ways. The next time you see a plastic bag caught in a tree or a gutter full of plastic bottles, straws, and lids, you might ask yourself where is this stuff going?

Captain Charles Moore of the Algalita Marine Research Foundation, who has been traveling and documenting the plastic flowing into our oceans, says source reduction is the only answer. He's written a book about it.[46]

I have been documenting the "stuff" in our lives for over twenty years, the things we use once and throw away. What all of us do, day to day, really matters. Recycling counts. So does remembering to bring your own bag to the store and saying no to single-use plastic when you do not need it.

My job as an artist is to transform the materials I find around me. This show at the Turchin Center has been my biggest opportunity yet. Thank you all for helping to collect and install and, most importantly, thank you for looking, seeing, and asking your own questions. The world belongs to all of us.

45. STREAMING: NEW ART OUT OF OLD BOTTLES

September 27, 2012

My show, *STREAMING: New Art Out of Old Bottles* opened today at the Gregg Museum of Art & Design at North Carolina State University in Raleigh. This show will be up until mid-December.

Below is the statement I wrote for the show:

> Most plastic is used only once, yet lasts forever. In the past sixty years, we have used single-use plastic with abandon. It is an integral part of our everyday lives. Try living for a day or a week or a year without it.
>
> For the past decade, my work has been looking at what we throw away in the United States. Americans create more garbage per capita than any other country. This was true twenty years

> ago, and it is still true now. Why is this? The answers are complex, and I believe that awareness is the first step in taking care of the world around us.

I made *STREAMING: New Art Out of Old Bottles* in collaboration with over 150 students from N. C. State's Arts Village, University Scholars, and College of Natural Resources, as well as museum staff and members of the greater community. All who came brought plastic bottles to the project. The students of Appalachian State University cleaned and de-labeled over 4,000 of the bottles you reused from a previous installation. This work is about community, process, reuse; and looking forward.

It contains:

1. A waterfall of plastic bottles flowing into a river of bottle caps and marine detritus.
2. A bamboo forest of green soda bottles and used chopsticks standing in a ground of bottle caps, lids, corks, and marine detritus.
3. A mountain range of over 3,000 PET plastic bottles.
4. A thundercloud of lids and straws.
5. Two red wolves made out of plastic bags, scraps of fabric, and string, wrapped over metal armatures.
6. Over 50,000 bottle caps and close to 5,000 plastic bottles.

All materials in this exhibition have had a previous life.

This work was a community endeavor, made possible by the hard work of many volunteers.

Heartfelt thanks to you all.

Tim Barkley

46. IF I HAD WINGS

May 8, 2013

> *The planet does not need more successful people. But it does desperately need more peacemakers, healers, restorers, storytellers, and lovers of every kind.*
>
> —DAVID ORR, *Earth in Mind*

It is a sunny morning on the North Carolina coast. I hear the sound of the ocean, and I hear birds; birds chirping, birds peeping, birds trilling, birds squawking, birds communicating in all sorts of ways. The waves crash, and the birds sing. When I arrived on Monday afternoon, it seemed like all the pelicans on the East Coast were flying overhead. I like to think it was in welcome, but know it's more likely there must have been good fishing somewhere close by.

I came here to rest after a hectic spring full of residencies all over North Carolina, as well as installations and a few shows. It has been a very busy year so far, and I am fortunate.

This morning, when I went down to take the first load of stuff to my car, I saw a red cardinal atop a telephone pole at the edge at the parking lot. Now, the mockingbirds are fighting the grackles at the bird feeders below. Crows and black-headed gulls are flying reconnaissance, and small long-legged seabirds are running in and out of the surf.

Why do we watch birds? For solace? Out of curiosity? For sheer delight?

I think one of the reasons I watch them is because they offer a window into another world, where politics has no meaning. A non-verbal world of weather and wings.

For the past few years, I have been making the birds I see out of stuff I find. Paradoxically, many of these birds are filled with single-use plastic bags. Next, they are wrapped with scraps of cloth and string. Finally, bits of found plastic, twigs, bark, feathers, or whatever might bring the bird to life.

To me, my birds are meditations on the natural world. Making them brings me some of the same peace I find in observing wild things. Solace, back to solace. And joy.

Currently, my birds are on display at Bull City Arts Collaborative in Durham as part of my show, *If I Had Wings*.

I wrote this statement for that show:

> Whoever we are, wherever we live—birds are wild and all around us. Pigeons and red-tailed hawks inhabit New York City. Crows are everywhere. Can we live among wild things and not dream of their wildness, their ability to fly above us and live beside us in places we do not know? Like many who live in an urban neighborhood full of trees, I feed the backyard birds. Daily, I watch red cardinals and black-white-and-gray chickadees as they gather with sparrows and wrens on my bird feeder. On the ground are pigeons, juncos, and an occasional rufous-sided towhee, scratching for fallen seeds. Most mornings, when I open my front door to empty the trash or go to my studio, I hear the crows calling from far above. As I write this, I watch a female cardinal with a bright orange beak and subtle green and brown feathers forage for seeds on the Rose of Sharon bush outside my office window. I watch the birds, and I wish I could fly . . .

To learn more about my wild animals, check out Danielle Maestretti's blog post about them for the American Craft Council.[47]

47. MUSHROOMS

June 6, 2013

You never change things by fighting the existing reality.
To change something, build a new model that makes the existing model obsolete.
—ATTRIBUTED TO BUCKMINSTER FULLER

Find a problem, not an idea. Then solve the problem.
—BURT SWERSEY, professor, Rensselaer Polytecnic Institute

Since late February, my world has been a rapid stream of residencies, workshops, shows, and installations. I have made recycled fish with first graders, wild animals with fifth graders and high school students, and a two-story "waterfall" with the help of student volunteers at the Cary Arts Center in Cary, North Carolina. I crossed the state, delivering my own animal sculptures to

Blue Spiral 1 in Asheville and Bull City Arts Collaborative in Durham, and I worked with Duke University students on an art installation about bioplastics. I have seen art using plastic and about plastic by other artists dealing with issues similar to my own. If you are in Chapel Hill this summer, be sure to check out Bright Ugochukwu Eke's work, made out of plastic bottles, at UNC's FedEx Global Education Center.

I have been traveling furiously down the river that is my life. Two weeks ago, I rode over to Raleigh with a bunch of interested citizens to visit the single-stream recycling facility. This is where all of the materials collected from our blue recycling bins are dumped, sorted, and processed. Seeing masses of everyday packaging flowing by on conveyor belts as people and machines separated stuff by type was like watching a river of consumption. It swirled and eddied and finally ended up in a bale, a box, or as refuse on the floor.

I write this post on a very gray and rainy day. Here in North Carolina, we are experiencing the remnants of Tropical Storm Andrea. The ground is soaked and rivers are filled to capacity, yet it rains on. Whatever we do, wherever we are, we are affected by the natural world around us. We cannot help it. For some of us, perhaps it is only when the weather gets very wet, or very stormy, or very dry, or "very" anything that we begin to pay attention. I have been running from job to job this spring, on the road, and in schools, art centers, and galleries. In the back of my mind, I always watch single-use plastic. I still try not to use it; I bring my own bags, say no to straws, and that

sort of thing. From the beginning of this blog, that has been my job, and it has not changed.

Plastics, with polystyrene at the forefront, came into common use during World War II. Since then, we have not looked back. No one at that time could have imagined the multitude of uses that would be developed for all sorts of plastics. Just as surely, no one could have imagined what making cheap, single-use but ever-lasting plastic items would cost us.

Why is this post entitled "Mushrooms?" If you read Ian Frazier's article in the May 20, 2013, *New Yorker* magazine entitled "Form and Fungus: Can Mushrooms Help Us Get Rid of Styrofoam?"[48] you will see why. Frazier tells the story of two young men, Gavin McIntyre and Eben Bayer, who founded a company called Ecovative Design. They develop ways to make polymers out of natural, biodegradable materials: mushroom spores. Their products are successful and compostable. Interest is worldwide. If you want to know more, you should watch their TED talk. Maybe finally we can develop a new system to replace single-use, downcyclable-at-best plastic. Such is my hope. Working harder and faster just keeps us working harder and faster, with no time to consider the results of our actions. We are not going to stop using plastic. Maybe we can learn to use natural polymers for a more sustainable world. What a thought!

48. PURIFIED: A RIVER IN THE DESERT

June 19, 2013

Saturday evening. I returned home from Odessa, Texas. I spent last week constructing my newest show, *PURIFIED: A River in the Desert.* This installation, at the Ellen Noel Art Museum, is made of 10,000 plastic water bottles. They were collected by the museum's small and dedicated staff, led by Curator of Education Doylene Land. For a mostly East-Coast girl, the terrain around Odessa and its close neighbor Midland was a new world to me. Both towns are in the Permian Basin, home of the largest inland petrochemical complex in the United States. This is where we get the fuel to drive our cars and live our lives. It is flat desert land, and the countryside is covered with low-growing mesquite. The climate is hot and dry, with big skies and grand sunsets. In town, close to town, and out of town, you can see rhythmic pumpjacks working constantly to draw oil out of the ground.

The Permian Basin has had very little rainfall in recent years. Oil is booming, but water is scarce. Water restrictions are a given. Unless the water is filtered, most people do not drink it; most everyone drinks bottled water out of necessity.

Last week, with the help of twelve smart and thoughtful high school students, we used all of the 10,000 bottles collected by the museum to make a waterfall flowing into a river. Everyone worked all week to make it happen, and we were well-rewarded for our efforts. PURIFIED, the first word of the title, came from the fine print on many of the water bottle labels; this was brought to my attention by one of the students. The work was a community endeavor. The museum has a staff of seven people, and everyone began bringing in their water bottles in February. A woman who owns a local maid service heard about the project and began to collect bottles from the houses she cleaned. Other interested citizens contributed as well. The gathering of the materials for this work of art was the foundation from which it grew.

49. HOME

September 1, 2013

Home is where the heart is.

—PLINY THE ELDER

People say that what we're all seeking is a meaning for life. I don't think that's what we're really seeking. I think that what we're seeking is an experience of being alive.

—JOSEPH CAMPBELL, *The Power of Myth*

We must let go of the life we have planned so as to accept the one that is waiting for us.

—ATTRIBUTED TO JOSEPH CAMPBELL

Earlier in the week, I drove with some new American friends over many kilometers. We went up and down, in and out, and roundabout through the rural farmlands of southern France. We were on our way from Auvillar, a

village on the Garonne River where I had been living for the past week or so, to drop one of them off at a point on the pilgrimage route, Le Chemin de St. Jacques. From there, she would walk for several days. We passed field upon field of sunflowers, soon to be harvested and made into oil. Occasionally, the roads were straight and lined with huge sycamore trees, planted by Napoleon during the French Revolution to keep his troops cool as they marched along. Sometimes, the roads were old and patched. Mostly, they were curved. Again and again, we passed fields, haystacks, and tiny villages on hilltops, each with a steeple pointing high in the sky.

The trip was a big adventure. It has left me grateful for the experience, as well as pondering the meaning of home. Besides being a physical location, I think home is a place I carry within me, in my heart, the very core of who I am. At present, I have left the town, the friends, and the house itself, all that I call home in a physical sense, to visit a new world a continent away. I am in the midst of a language I do not know, a place I have never been. I am living next to the Garonne River, in Auvillar. The church a few buildings down the road from me was built in the year 900 C.E. History here is palpable.

What do I know about where I am? I have seen more sunflowers than I could ever imagine. I have eaten strawberries sweeter than any I have ever known. The cheeses I have bought—chèvre, Roquefort, and so many more—are food for the gods, as is the local honey. Bees are everywhere in the flowers that fill the streets.

I suppose it's not surprising but, just like back home, the people that I've met in France recycle and are concerned about the environment. I hear conversations about how different fruits are late or early, depending upon how they are affected by the changing weather. Another example from Europe: In one of my favorite mystery series (set in Venice), the young daughter of the main character counts gas miles and eschews water drunk out of plastic bottles.

We carry our sense of home, our love of it, our longing for it, with us all the time. Everywhere. One world, a big world, our world; it is our home.

Many thanks to the people of VCCA France, the town of Auvillar, the artists, the merchants, the travelers, the new friends made, and the kind people on the streets for loving where they live and sharing it with those of us who are traveling through, on the *chemin* or for a longer time.

50. WHAT IS MR. DUKE THINKING?

September 13, 2013

Last Friday, on the campus of Duke University, I made a superhero cape for James B. Duke himself. This happened while students built *Fort Duke* out of used move-in boxes in a successful attempt to break a world record. All of this was a big kickoff for Duke's Fall Arts Festival and this year's focus on sustainability. We worked on the quad in front of the Duke Chapel, where a statue of James Buchanan Duke presides. He has a cane in one hand, a cigar in the other, and stands atop a tall stone plinth. Under a cloudy sky, over 250 students, staff, and faculty worked all day to make a construction out of over 3,500 cardboard boxes. Architect and Duke instructor Todd Berreth designed a structure that was both a fort and a maze. From the beginning, everyone worked together like bees in a hive. As the day progressed, the hum of activity grew in intensity as the dimensions of the fort increased. Box upon box was assembled, numbered, and stacked. The walls grew higher. The paths to the center grew deeper, longer, and more convoluted. Around six in the evening, a huge cheer rumbled across the quad as the record-breaking box was placed on the fort. It was a race against time and weather, with everyone on the same team. In the end, the fort was a joy to behold. Arwen Buchholz of Duke Recycles told me that it was all unstacked, crushed, and recycled before eleven that same evening. Teamwork at its best! That's for sure.

It was clear that Mr. Duke, who was standing tall in the center of all of this activity, should be included in the action. So, under my direction, we made him a shining twenty-foot-long cape of plastic water bottles. Many of the bottles said, "Welcome Duke Students, Class of 2017."

Early in the day, two graduate students from Duke's Pratt School of Engineering came by to lend me a hand. One of them asked, "So, why are you putting a cape of plastic bottles on Mr. Duke?"

Good question, I thought. "It is a recycled cape. He is our superhero today."

"So, what is Mr. Duke thinking?" Another excellent question.

I answered, "Mr. Duke is thinking about sustainability."

James B. Duke was a magnate of tobacco and textiles. He established the Duke Endowment in 1924, continuing a legacy of giving begun by his father, George Washington Duke. He was a champion of industry and believed in sharing his wealth for the greater good. I like to think that if he could have lived another lifetime, he would have put out his cigar and been an ardent recycler. Like the university endowed by members of the Duke family, he would look for solutions to make the world sustainable.

In the words of the naturalist Edward O. Wilson: "The great challenge of the twenty-first century is to raise people everywhere to a decent standard of living while preserving as much of the rest of life as possible."[49]

Sustainability. It concerns us all, yet how much do we really think about it? Last Friday, over 250 students, faculty, and staff were doing just that as they worked together. I can't help but hope that all of us involved will continue to do so. The fort and the cape are gone, yet the community created by their making remains as inspiration. Sustainability. It is the biggest challenge we face as we live together on this earth. We have a lot of work to do. Paying attention and working together are key factors here. Building Fort Duke was fun. All day, ever-changing groups of people worked long and hard to make it happen. The mantra "we can do this" was a palpable undercurrent.

51. THE FOX

January 29, 2014

> *The size of the place that one becomes a member of is limited only by the size of one's heart.*
>
> —GARY SNYDER, *Back on the Fire*

> *Though both the red fox and the gray fox live in North Carolina today, the gray fox is the state's only native fox species. . . . The gray fox is slightly smaller than the red fox and is much darker in overall coloration. . . . The overall coloration is best described as a salt and pepper gray with a dark streak extending down the back, along the top of the tail.*
>
> —NORTH CAROLINA WILDLIFE RESOURCES COMMISSION

A gray fox (*urocyon cinereoargenteus)* is buried in my backyard. I live in a city neighborhood with lots of oak trees and a stream at the bottom of a hill. Ellerbee Creek flows along a bike path that ends at the golf course. A while back, I used to see a great blue heron wading there. A bit of forest remains between the last row of houses and the stream. Across this slice of water, you can see the backs of all the businesses that line Guess Road, a busy four-lane road that crosses under I-85 and is the home of many gas stations, convenience stores, motels, hotels, and pawn shops. I heard for a while that neighbors were spotting a gray fox around there, but I never saw him.

But then a fox came to me. Last November, right after our local elections, my neighbor and newly chosen councilman Don Moffitt called me on the phone one afternoon. "Bryant," he said, "I was out by the lemur center when I saw an animal by the side of the road. I stopped, because I thought it might have been an injured lemur, but it was a fox. Would you like to see him?" Well, of course I wanted to see him. Don knows of my interest in wildlife and how, when I can find the time, I work in my backyard studio making the animals I see around me out of recycled materials. He told me that he, his daughter, and her friend would be right over. The previous Thanksgiving, Don and I, along with other friends who were sharing the holiday together, took a walk after dinner and ended up at my studio, where we looked at my attempt at making raccoons out of recycled materials. Earlier, we had all admired the gray fox that stood on the credenza in our hosts' dining room. They had come across it by the side of the road while biking out near Hillsborough a year past, and our friend Susan insisted that they take it to a taxidermist. Truly, it was a thing of beauty, with its beady glass eyes staring down at us as we ate our turkey. Clearly, Don and I had foxes in our history together. Somewhere in the conversation about the fox Don had found I asked, "Would you like to bury him in my backyard?" Soon he arrived with a shovel and a fox wrapped in a blue tarp. The two girls were close in attendance. Quickly, he dug a hole for the fox. After he nestled the dead animal into the ground and re-covered it with tamped-down earth, we placed a large garden stone over the grave.

My question is this: Do we have room in our hearts for wild things? I am glad I live in a world where foxes forage, where rabbits occasionally show up in my garden, and where birds fly above me. I know I am fortunate to have friends who care about these things as well.

Foxes are predators. They catch small rodents and eat them. Then they sleep and mate and do it again. They do not go to the dentist and are never late. They do not play soccer or drive a car or do anything remotely human, and we do not want them to. We want them to be foxes. They are the other. We cannot know them. We can only be aware of their unfathomable existence in our lives. Acknowledging the other makes our world bigger, deeper and infinitely more complex.

It takes us out of ourselves. And when we return, we are better for it.

North Carolina Wildlife says that fox sightings are increasingly common across North Carolina, either because of the abundance of food available to foxes or because they may have been disturbed from their resting places.

Jen Crickenberger

52. MARKERS OF OUR LIVES

March 15, 2014

I have a lot of bottle caps and jar lids. I stopped attempting to keep track at around 100,000. Friends and family collected them for me for about ten years; I stopped when I could not store any more. Clearly, I had enough to tell the story, to paint the picture.

I spent last week at the Cornelius Arts Center in Cornelius, North Carolina, working with the community to make a mandala out of some of the caps and lids. Over the past decade, I have used and reused these caps to make other mandalas. That said, each time is a new time, each community coming together in its own unique way.

The mandala process begins at my home, where I drag out boxes and bins of caps from under tables, from my front porch, and from the shed down the street. I pick and sort which colors and types of lids to bring. When I arrived last Tuesday, Jen Crickenberger and her staff quickly and efficiently unloaded my truck. We conferred about traffic flow in the gallery and what her thoughts on the size and form of the mandala might be. Within twenty minutes of my arrival we were at work, measuring and gridding the space and just as quickly counting and stacking caps. Volunteers began dropping in to help. Shay, a scientist who had recently immigrated from Nigeria, came every afternoon like clockwork. On Thursday, we had students from two different elementary schools in the area working on the mandala.

These mandalas are projects of many pieces, containing from 10,000 to 50,000 lids and caps. Coke caps, pickle lids, beer caps, and plastic caps in many colors from milk bottles. Lids from juice bottles, lids from salsa, applesauce, peanut butter, and jam jars. The lids truly are markers of our lives. They are records of consumption. They represent sustenance, yet sustenance that is processed, one step removed from the ground from which it came. The first glass canning jars were invented in 1858, coincidentally the same year as the can opener was patented by Ezra Warner of Waterbury, Connecticut. Both of these items marked the beginning of mass processing and storage of food, and we have not looked back for a minute since—as the caps in my mandalas confirm. Each cap is a mark of something packaged, purchased, and then consumed.

Many people think that a lot of folks helped me in collecting these caps, but that is not necessarily true. It was the consistency of collection that gave me the quantity I now have. Personally, I buy a glass container of milk every week, each one sealed by a plastic cap. Fifty-two navy-blue milk lids per year, which multiplies to 520 in ten years. One person, one item, for ten years. Those blue caps are only the tip of my own personal iceberg of consumption. Like Friday followed Robinson Crusoe on his island, my carbon footprint follows me.

Markers of Our Lives: A Community Mandala was on display at the Cornelius Arts Center through April 30, 2014. Special thanks goes to Jen Crickenberger and her hard-working staff of Jake, Nicole, and Suzanne. Thanks also to the town of Cornelius for believing in this project and the community for helping to make it.

You can watch the week unfold in thirty-nine seconds in a video made by Jen Crickenberger![50]

Jeanette Stokes

53. THE VIEW FROM PELICAN HOUSE

January 8, 2015

Some Thoughts on Beauty—
Shells, all kinds, broken, whole, barnacled—all of them
The ocean and how it never fails to open my heart, wave upon wave
A basketball pass completed from one running girl to another across the entire court
The moon
A mackerel sky, a buttermilk sky, almost any sunset
Seeds first coming up
Frost on the windshield with the sun shining through
The Eno River, anywhere
Many memories of days long gone
The fleeting present all the time

—BRYANT HOLSENBECK, *January 2015, Salter Path, North Carolina*

I started *The Last Straw* on January 1, 2010. Five years ago, I began to do all that I could to eliminate single-use, one-time-only, toss-away plastic from my life. Using Barbara Kingsolver's *Animal, Vegetable, Miracle* as inspiration, I began my journey. Kingsolver's quest to eat locally for a year meant eating seasonally and relying on local farm produce. Each family member got to pick one thing as an exception, such as coffee or chocolate.

Being single, I allowed myself up to four dispensations. I chose my contact lenses, their packaging, and the cleanser involved, and the prescription drug I take for migraines, which is non-generic and comes in a non-biodegradable blister pack. Everything else, from chocolate chips to yogurt, to takeout food, I find a way to make, I buy without plastic packaging, or I do without. Often, I thought I might give up, but I always found a solution, another way. Many people helped me on the journey. My yoga group cheered me on and did their own experiments trying to use less plastic. To this day, I think everyone in that group carries a refillable water bottle and brings their own bags to the grocery store.

For a full year, I did as I intended as best I could. I baked, I cooked, I shopped, and I discovered ways to live without. In the four years since, I have not been as strict with myself, but as I look back I see clearly how that year changed my habits:

- I always think, "Do I need that bag?" when I purchase something. Bringing bags into the grocery store or farmers' market is a given. If I forget, which is not very often, I put stuff in my purse, carry it out in my hands, or go back to the car. I have trained myself like a housebroken dog. I know what to do, and except for very rare occasions, I do it.
- Eating out, I look for places that do not use plastic utensils or Styrofoam containers. In Durham, there are several restaurants I just avoid. I am fortunate to live in a town with lots of great choices. A few years ago, I became a founding member of "Don't Waste Durham," a group of individuals looking for ways to permanently

Tim Barkley

reduce the use of Styrofoam in our schools and restaurants. Several members are working hard to develop composting for restaurants and homes. One of the initiatives on the table is reusable takeout containers for restaurants. Stay tuned!

- I have become a constant collector of plastic along the coastlines. I accumulate this everyday detritus whenever I am near large bodies of water, and generous friends collect for me as well. One of my yoga partners, Grace, spends time on both the Atlantic and Pacific coasts, with an occasional Caribbean island thrown in as well, and she collects for me wherever she goes. I have begun using these materials to make animals, mostly birds, currently chickens.

I continue to marvel at our overuse of plastic water bottles, especially here in the United States, where the water is predominately clean. When did we decide it was better to buy water, processed and formulated and put into a plastic bottle? In this vein, my work as a public artist has grown. With help from community organizations and interested citizens, I have made a number of plastic bottle waterfalls. The largest of these, *PURIFIED: A River in the Desert,* was in the Ellen Noel Art Museum in Odessa, Texas, and contained over 10,000 bottles collected by the museum staff and members. This fall

I worked with Theatre Art Galleries and High Point University students to install a three-story waterfall in the High Point Public Library. I also hung a water-bottle waterfall in Winston-Salem, North Carolina, as part of a pop-up art series sponsored by Piedmont Craftsmen. Two similar installations are planned for spring 2015, and future work along these lines has been planned.

Currently, I am spending the first week of 2015 at a writer's retreat on the edge of the Atlantic Ocean. The waves are a constant rhythm, day and night. All meals are cooked for us and served on real plates, using real silverware, with the good company of six other writers.

The beach is clean and wide, and I am happy to report that I have not found much plastic washed ashore or deposited on the edge. I have a small bag of stuff collected—a rubber glove, two or three pieces of plastic sheeting, four or five bottle caps, one balloon, and maybe ten synthetic ends of cigarette butts, the filters washed white from the surf and sand.

I have a job to do. Here is what I think it is: to continue documenting, collecting, and making art out of the detritus we humans create in the world and to live my life with love and appreciation for the beauty and many kindnesses I find here on earth.

ACKNOWLEDGMENTS

I want to thank the many people who encouraged me along the way. Your support helped me to keep going, to keep paying attention, and to turn my many frustrations into new ideas and ways of seeing.

I remain deeply grateful to the women in my Wednesday yoga group who tried out my ideas and encouraged me throughout my trial year of living without disposable plastic.

Miriam Sauls, Sarah Bingham, and Terry Vance provided wonderful editing support while I was writing my blog. Thank you!

I deeply appreciate Jeanette Stokes, Anita McLeod, and the Resource Center for Women and Ministry in the South for believing in my art and activism, helping me to keep going, and encouraging me to share my ideas about environmental stewardship with the wider public. Also, Marcy Litle, Liz Dowling-Sendor, and Bonnie Campbell for all the work they did to help my blog become this book.

Finally, I want to thank everyone who believed in me when I decided to try living without single-use plastic and gave me ideas to keep me going when I was stuck.

You have all helped me to see that what we do each day truly matters. Deep gratitude to you all. All the time.

NOTES

1. Thich Nhat Hanh, *The Heart of Understanding: Commentaries on the Prajnaparamita Heart Sutra,* 2nd ed. (Berkeley, CA: Parallax Press, 2009), 3.
2. U.S. National Park Service, Mote Marine Laboratory, Sarasota, FL, http://des.nh.gov/organization/divisions/water/wmb/coastal/trash/documents/marine_debris.pdf.
3. "Litter Facts," NC Public Safety, http://www.ncdps.gov/DPS-Services/Crime-Prevention/Litter-Free-NC/Litter-Facts.
4. William McDonough and Michael Braungart, *Cradle to Cradle: Remaking the Way We Make Things* (New York: North Point Press, 2002).
5. For more, see her blog. Rebecca Currie, *Less Is Enough* (blog), https://lessisenough.wordpress.com.
6. "Songs of Hope for Haiti," *Partners in Health,* https://donate.pih.org/page/outreach/view/personal/SongsOfHopeForHaiti.
7. Currie, *Less Is Enough.*
8. For more information, check out their website: www.weaverstreetmarket.coop.
9. Barbara Kingsolver, *Animal, Vegetable, Miracle* (New York: Harper, 2007).
10. Joe Dickson, "The FDA Changes Its Tune on Bisphenol-A," *The Official Whole Foods Market Blog,* January 19, 2010, https://www.wholefoodsmarket.com/blog/whole-story/fda-changes-its-tune-bisphenol.
11. Currie, *Less Is Enough.*
12. Joe Graedon, "How Hazardous Are Plastic Containers?" *The People's Pharmacy* (blog), April 13, 2008, http://www.peoplespharmacy.com/2008/04/13/how-hazardous-a/.
13. Kate Greenstreet, www.kickingwind.com.
14. "Plastic," *Merriam-Webster,* accessed February 26, 2010, http://www.merriam-webster.com/dictionary/plastic.
15. T. S. Eliot, "The Hollow Men," *The Complete Poems and Plays* (Boston: Houghton, Mifflin, Harcourt, 2014), 56–59.
16. Bryan Walsh, "The Perils of Plastic," *Time,* April 1, 2010, content.time.com/time/specials/packages/article/0,28804,1976909_1976908,00.html.
17. *The Story of Bottled Water,* directed by Louis Fox (2010: Free Range Studios), storyofstuff.org/movies/story-of-bottled-water/.

18. *Waste Land*, directed by Lucy Walker (2010; New York: Arthouse Films, 2011), DVD.
19. *Addicted to Plastic*, directed by Ian Connacher (2008; Toronto: Cryptic Moth, 2009), DVD.
20. Mike Melia, "First, the Great Pacific Garbage Patch; now the Great Atlantic Patch," *The Toronto Star*, April 25, 2010, www.thestar.com/news/world/2010/04/15/first_the_great_pacific_garbage_patch_now_the_great_atlantic_patch.html.
21. "Duke Leaf Award," *Duke Nicholas School of the Environment*, https://nicholas.duke.edu/events/duke-leaf-award.
22. McDonough and Braungart, *Cradle to Cradle*.
23. For more on Safe Passage, check out their website: safepassage.org.
24. *Camino Seguro* website, accessed June 17, 2010, http://www.safepassage.org/.
25. Ibid.
26. *PLASTIKi*, http://theplastiki.com.
27. Used by permission of the author.
28. *No Impact Man*, directed by Laura Gabbert and Justin Schein (2009; New York: Oscilloscope Laboratories, 2010), DVD.
29. "The Unplugged Challenge," *The New York Times*, August 15, 2010, http://www.nytimes.com/interactive/2010/08/02/technology/unplugged.html?_r=0.
30. Annie Berthold-Bond, *Better Basics for the Home: Simple Solutions for Less Toxic Living* (New York: Three Rivers Press, 1999), 203.
31. "About Deep Roots Market," *ATH, Practitioners, Organizations*, http://www.allthingshealing.com/athforum/profile.php?uid=927.
32. *The Majestic Plastic Bag—A Mockumentary*, directed by Jeremy Konner (Santa Monica, California: Heal the Bay, 2010), http://www.healthebay.org/media-center/videos.
33. Alden Wicker, "Seven Ways to Kick the Plastic Habit: Tips and Tricks for Living Plastic Free," *Huff Post Green*, September 14, 2010, http://www.huffingtonpost.com/2010/09/14/7-ways-to-kick-the-plasti_n_710987.html.
34. Chris Jordan, *Running the Numbers, an American Self-Portrait*, http://www.chrisjordan.com/gallery/rtn/#silent-spring.
35. Dianna Cohen, "Tough Truths about Plastic Pollution," *YouTube*, https://www.youtube.com/watch?v=fddYApFEWfY.
36. Salman Rushdie, *Luke and the Fire of Life* (New York: Random House, 2011).
37. "What Does Technology Want?" *Radiolab*, November 16, 2010, http://www.radiolab.org/story/101024-idea-time-come/.

38. Steven Johnson, *Where Good Ideas Come From: The Natural History of Innovation* (New York: Riverhead Books, 2010).
39. Kevin Kelly, *What Technology Wants* (New York: Penguin, 2011).
40. *Bag It*, directed by Suzan Beraza (New York: Docurama Films, 2010).
41. *The Monticello Dialogues: William McDonough in Conversation with Michael Toms* (New Dimensions Foundation, 2004).
42. Jalal al-Din Rumi and Coleman Barks, *The Essential Rumi* (New York: Harper Collins, 1995), 36.
43. R. Buckminster Fuller, *Critical Path* (New York: St. Martin's Press, 1981), xxxviii.
44. "Be Straw Free Campaign," *ecocycle, Building Zero Waste Communities*, http://www.ecocycle.org/bestrawfree.
45. Lyanda Lynn Haupt, *Crow Planet: Essential Wisdom from the Urban Wilderness* (Boston: Little, Brown, 2009), 7.
46. Charles Moore, *Plastic Ocean* (New York: Penguin, 2011).
47. Danielle Maestretti, "Bryant Holsenbeck's Wild Recycled Animals," *American Craft Council* (blog), January 18, 2013, https://craftcouncil.org/post/bryant-holsenbecks-wild-recycled-animals.
48. Ian Frazier, "Form and Fungus: Can Mushrooms Help Us Get Rid of Styrofoam?" *The New Yorker,* May 20, 2013, http://www.newyorker.com/magazine/2013/05/20/form-and-fungus.
49. Edward O. Wilson, *The Creation: An Appeal to Save Life on Earth* (New York: W. W. Norton, 2006), 6.
50. "Three Days and Thousands of Caps Later, It Is Done," *Cornelius Arts Center Facebook Post*, March 7, 2014, https://www.facebook.com/photo.php?v=841580745857579&set=vb.118534851495509&type=2&theater.